FROM ONE CELL

"Dr. [Ben] Stanger artfully guides us through key experiments that contributed to the foundations of our knowledge about embryonic development." —Adrian Woolfson, *Wall Street Journal*

"Remarkably effective. . . . *From One Cell* is replete with fascinating stories." —Jonathan Shaw, *Harvard* magazine

"An authoritative account of a critical area of medical research and the promises it holds." —*Kirkus Reviews*

"Fascinating from beginning to end, Ben Stanger's *From One Cell* takes us through the most important journey in all of our lives. Stanger is an enthusiastic and knowledgeable tour guide through the science that will impact our lives in the future."

—Neil Shubin, paleontologist and author of *Your Inner Fish*

"An inspiring, masterful, and authoritative account of [a] vital scientific frontier—rendered in brilliant, beautiful prose—*From One Cell* is about you, your beginning, and your future." —Daniel M. Davis, author of *The Beautiful Cure* and *The Secret Body*

"*From One Cell* is a pleasure to read—elegant, accessible, and endlessly exciting."

—Robert A. Weinberg,
Massachusetts Institute of Technology,
author of *The Biology of Cancer*

"One of the great mysteries in all of science is how individual fertilized eggs transform into the beautiful, exceedingly complex forms of life we see in nature and in ourselves. . . . Lucid and engrossing, *From One Cell* is a wondrous journey." —Cliff Tabin, professor and chair of genetics, Harvard Medical School

"Remarkable. . . . I picked up the book already familiar with many of these tales, but Stanger brings them and the scientists to such vivid life that I could hardly put it down. A fascinating read."

—Tony Wynshaw-Boris, Case Western Reserve University School of Medicine

"If you're even the tiniest bit interested in how all cells in our body came from a single cell, I highly recommend that you read this book."

—William Anderson, director of education, Harvard Department of Stem Cell and Regenerative Biology

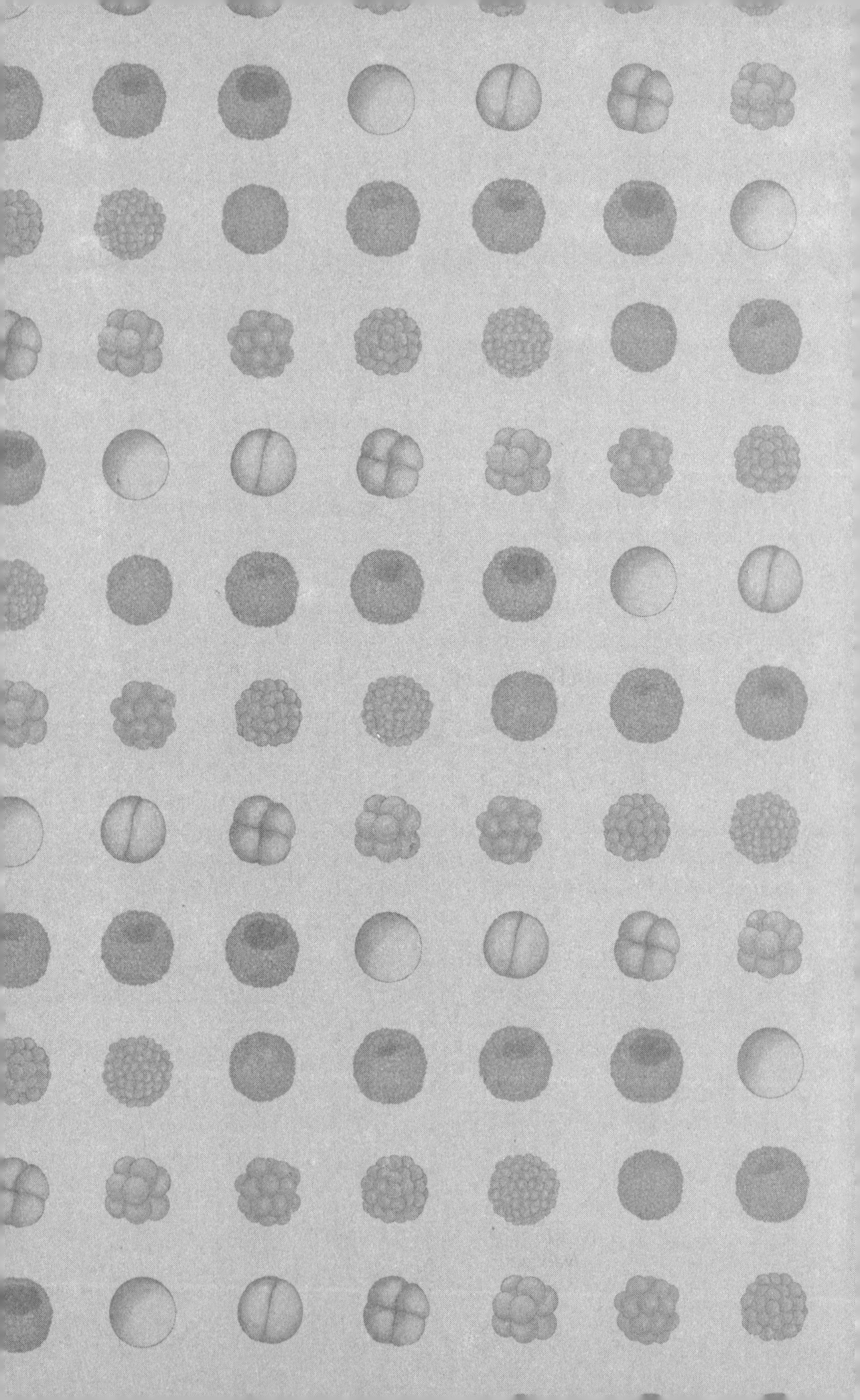

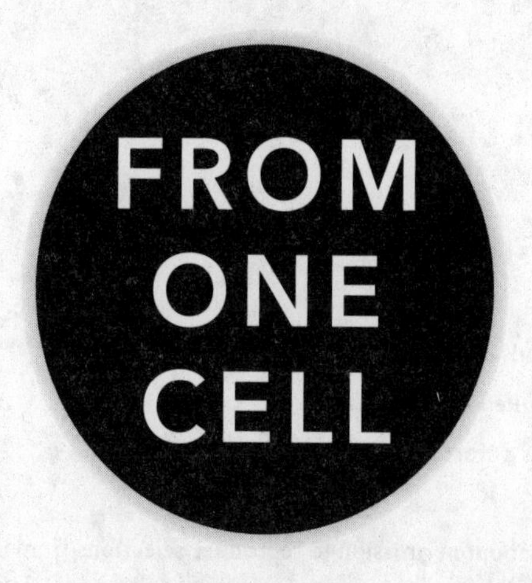

FROM ONE CELL

A Journey into Life's Origins and the Future of Medicine

BEN STANGER

W. W. NORTON & COMPANY
Independent Publishers Since 1923

Printed in the United States of America

First published as a Norton paperback 2024

For information about permission to reproduce selections from this book, write to Permissions, W. W. Norton & Company, Inc., 500 Fifth Avenue, New York, NY 10110

For information about special discounts for bulk purchases, please contact W. W. Norton Special Sales at specialsales@wwnorton.com or 800-233-4830

Manufacturing by Lakeside Book Company

Book design by Daniel Lagin

Production manager: Lauren Abbate

ISBN: 978-1-324-07604-9 pbk.

W. W. Norton & Company, Inc.

500 Fifth Avenue, New York, N.Y. 10110

www.wwnorton.com

W. W. Norton & Company Ltd.

15 Carlisle Street, London W1D 3BS

1 2 3 4 5 6 7 8 9 0

To Beatrice and William

Freedom is not worth having if it does not include the freedom to make mistakes.

— ATTRIBUTED TO MAHATMA GANDHI

A thing constructed can only be loved after it is constructed; but a thing created is loved before it exists.

— G. K. CHESTERTON, *APPRECIATIONS AND CRITICISMS OF THE WORKS OF CHARLES DICKENS*

There's a paradox in every paradigm.

— ANI DIFRANCO, "PARADIGM"

CONTENTS

FROM ONE CELL

PROLOGUE: CONCEPTION

Before any of us laughed, felt pain, danced, underslept, overindulged, loved, lost, or took our first breath, we completed a journey. It was a trip that involved extraordinary growth and hefty loss, dramatic movement and eerie stillness, processes of unimaginable complexity and yet remarkable reproducibility. Our voyage grew out of a remote past, recapping millions of years of trial and error and genetic memory, and it will assuredly stretch into the future, as generations yet to come follow the same path we somehow managed to navigate. No matter how estranged we may have become from other members of our species through differences in ideology, language, or culture, our starting place was the same. We are bound together because of this common experience—from the simplest of beginnings, we emerged from the womb as human beings.

The need to comprehend our origins runs deep. Intuitively, we recognize that only a small fraction of what happens around us is observable. But this limited visibility does not prevent us from yearning to expose the unseen. The phenomena that lie beyond our day-to-day perceptions have an irresistible pull, whether they involve mapping the edge of the cosmos or probing the inner workings of the cell. Explanations that take us back to the source—cause-and-effect accounts of how things came to be the way they are—are particularly compelling for the way they bring order to an otherwise chaotic universe.

The fundamental truth of our origin is this: every animal on earth starts its life as a single cell. But how does all the information necessary to make something so complex get compressed into something so simple? How do the trillions of cells generated from this peculiar unit know what to become and where to go? How can a better understanding of our embryonic past help us lead a healthier future? These are the questions these pages will strive to answer.

This book relates a narrative that has been rehearsed countless times: the story of how a single cell grows into a mature organism. The journey begins with the fertilized egg (also known as the zygote), the basic unit of animal life, and traces how we arrived at our modern understanding of cells, genes, and embryology. We will explore the dueling contributions of nature and nurture—the balance between rigidity and pliability of fate necessary for normal development—to understand how a single cell can spawn the many cell types that make up our bodies. Along the way, we will examine the pivotal role that stem cells play in the process.

The story of the embryo is grounded in chemistry, as all the information needed to make an entire animal is stuffed into the DNA genome of the zygote. Our journey will thus explore how our understanding of genes grew from earliest concepts to the current state of our knowledge. We will scrutinize fundamental questions regarding genes and development: What happens to the genetic information that cells no longer need? How do cells overcome the "tyranny" of genes, enabling them to acquire specialized characteristics in the face of a uniform set of genetic instructions? How does "*epi*genetic regulation"—heritable changes that don't involve our DNA sequences—determine a cell's very identity?

In the end, we will consider where this science of the embryo is leading in terms of our health. The emerging field of regenerative medicine, based on our nascent understanding of how tissues are molded in the embryo, has the potential to revolutionize how we treat a variety of diseases and conditions. We will consider the unsettling similarities between embryos and tumors, and how studies of development are leading to new cancer therapies. We will examine the causes of organ failure and contemplate why some organs and organisms are capable of regenerating while others are not. And we will eavesdrop on the promising efforts that are underway to apply lessons from the embryo to medical practice. These include ominous-sounding processes like "cellular reprogramming" and "gene editing" that have the power to create artificial tissues or rewrite our genomes.

There is a danger in writing this sort of book. On the one hand, it risks venturing into territory that is overly technical. Each of the topics we cover—cellular differentiation, morphogenesis (the establishment of form), genetics, stem cell biology—could fill volumes, and it is hard to discuss molecular embryology without slipping into jargon. (This recalls the warning that physicist Stephen Hawking received while writing *A Brief History of Time* that each equation he included in the book would cut sales in half.) On the other hand, there is a risk of oversimplifying the science or worse—indulging in exaggeration that results in some fantastical image of a utopian or dystopian future.

Instead, we will tread that line by focusing on the foundational discoveries that built and continue to advance our knowledge. The title—*From One Cell*—refers to two different but related types of cells with extraordinary potential: the zygote, which gives rise to every cell in the body during embryonic development, and the embryonic stem cell, which has a similar capacity. It is common knowledge that scientists hope to use stem cells to treat a variety of diseases. But what is less widely appreciated is that most of what we hope to achieve with these cells, or their surrogates, the embryo accomplishes as a matter of routine. Thus, for regenerative medicine to deliver on its promise, our best bet is to look inward—to see how our bodies, and those of our phylogenetic predecessors, were originally fashioned. In short, we need to understand normal development.

After more than two decades of work in the field—as a physician, developmental biologist, and now cancer biologist—I am still awed by the beauty, mystery, and universality of the embryo. Yet as a practicing doctor, I also view the embryo through a clinical lens, given its potential to deliver remedies for myriad diseases. It is the nature of research that any advance raises dozens of new questions; indeed, this self-perpetuating knowledge gap is what makes science so irresistible. The story of embryonic development, stem cells, and regeneration is therefore still being written. Even so, it is a compelling account, involving scientists, philosophers, patients, and doctors. But the star of our narrative is the embryo—the entity from which all animals, including human beings, arise.

From One Cell tells the complex and miraculous story of the embryo—the story of how we come to life.

Chapter 1

THE ONE CELL PROBLEM

Which came first, the chicken or the egg?

— ANCIENT RIDDLE

Where do babies come from?

It is a question we have all pondered at some point in our lives, a question as personal as it is universal. There are, of course, many ways to answer this question, but the one that we most often learn in school is that life begins as a single cell, the product of fertilization between egg and sperm. That cell divides once, then again, and again, and so on, until a fully formed human being, containing more than a trillion cells, ultimately comes into existence.

To a first approximation, this description of the embryo's development, also called embryogenesis, is accurate. But if we step back for a moment and think about all the things that must go right during this process, it's a marvel that any of us are here. As the fertilized egg divides, its progeny coalesce into structures (organs) so complicated that we cannot begin to engineer them outside of the body. With every division, a cell makes a high-fidelity copy of its DNA, which it then passes along to its daughters. With time, those daughters, and their descendants, take on new identities—blood, bone, skin, and so on—that make them wholly distinct from their parents.

The end product of these cellular gymnastics is a body ready for countless contingencies—a scarcity of food and water or threats from predators, infectious agents, and toxins. There are countless opportunities for error, and yet mistakes are quite rare. Every day, billions of animals take the final step in a journey that led them from solitary cell to fully formed organism, a daily tally that includes over 300,000 human beings. Embryonic development is the primary and most important rite of passage, the supply chain for virtually every new animal on our planet.

How does the marvel of development happen so seamlessly and reproducibly? How do cells know how to specialize, when to divide, where to go, and how to behave? Is development principally controlled by our *genes* or by the environment? How does a species ensure that the process of development repeats itself with each new generation, limiting errors that might cause it to face extinction? And most remarkable of all, how does an entire animal—one capable of movement, respiration, digestion, sensation, and reason—arise from a single cell? It is a conundrum we can refer to as the "One Cell Problem."

Human beings have wrestled with versions of these questions since antiquity, but until the mid-nineteenth century we lacked the technology to observe embryos with any degree of refinement. Hence, most of the concepts about development that emerged before this time were incomplete, or flat-out wrong. With powerful molecular tools at our disposal, we now find ourselves at a unique moment—a time when we can point to the genes, cells, and molecules that orchestrate development and describe, in broad strokes, how we come into being.

TURTLES ALL THE WAY DOWN

LET'S BEGIN WITH A THOUGHT EXPERIMENT. IMAGINE FOR A MOMENT that you are in a state of ignorance about embryos, unaware of the existence of cells or genes. Instead, picture the human form as you perceive it day-to-day: as an unbroken frame composed of body parts, some large and some small. Knowing nothing about the body other than what you can see—head, trunk, limbs, eyes, ears, mouth, teeth, nails, hairs, and so on—try to imagine how it arose, tracing its origins in reverse time from adult to newborn to . . . whatever came before.

Personally, I find this a difficult exercise. The problem arises from having too much information—specifically, from knowing that the body is made up of cells. That simple piece of knowledge ties embryonic development to those microscopic units, the building blocks of tissues and bodies. Consequently, it is as challenging (for me) to "unknow" the fact that embryos are made up of cells as it would be to imagine a sand dune without comprehending that it is made up of barely visible grains. Learning is typically a one-way street.

To get around this conceptual hurdle, I turned to my six-year-old daughter, who I knew had not yet learned about such things in school.

"Sarah," I asked, "where do babies come from?"

Sarah was used to fielding these kinds of unusual questions—provocative scientific, philosophical, or metaphysical propositions—and she always played along, knowing how much her answers delighted me. (Even so, the irony that I was asking my daughter where babies come from, and not vice versa, did not escape me.)

"From Mommy's tummy," she answered.

"Yes, that's right," I said. "When a baby is still inside its mommy's tummy, it is called an embryo."

"Oh," she said, unimpressed. I continued.

"Now, Sarah, tell me what you think an embryo *looks like* when it is inside its mommy's tummy."

She thought about this for a second before responding. "It looks like a *tiny* baby, smaller than a chicken wing."

"Ah. Interesting," I said. "And what about before that? What does an embryo look like at the beginning, when it is just starting to grow, *before* it looks like a tiny baby?"

The answer seemed so obvious that she began to laugh.

"Before that, it looks like a *tiny tiny tiny* baby."

THE NOTION THAT WE BEGIN AS "PRE-EMBRYOS," SQUISHED-UP VERsions of the beings we will become, with all our parts in place from the moment of conception, is known as *preformationism*. The idea may seem silly and simplistic, but it is not farfetched at all, as my daughter's answer suggests. Indeed, the notion that animals start as minuscule preformed babies—like a photographic image placed under an enlarger—may be the most *intuitive* way to think about development, particularly if one is not saddled with notions of cells, genes, or evolution.

The early Greeks spent a good amount of time debating the nature of the embryo, and most sided with the concept of preformation. But in the fourth century BC, Aristotle weighed in. He reasoned that if the body's form indeed predates development—wholly intact from its earliest stages—then an observer watching the process should see its structures come into view all at once, a complete unit rather than a piecemeal assembly. To test this, the philosopher examined dozens of chick embryos at various stages of development. Noting that the chicken's heart appeared (and started to beat) well before other organs, Aristotle concluded, correctly, that the parts of the body come into being successively, not via the expansion of something preformed. The phenomenon became known as *epigenesis* (*epi*-, upon; *genesis*, origin) to reflect the idea that animals grow through a gradual assembly of parts upon parts. Aristotle's logic silenced the preformationists, and epigenesis emerged as the dominant paradigm for the next two millennia.

Paradoxically, it was the rise of microscopy in the mid-seventeenth century that fueled preformationism's return to favor. The first microscopes were invented by Antonie van Leeuwenhoek, a Dutch textile merchant who, looking for better ways to assess the quality of thread in a piece of cloth, invented a method for making lenses that could magnify objects beyond anything in existence. By 1670, his hobby of lens-grinding had become a full-time obsession. As he looked through his new microscopes, subjecting every specimen he could find to their refractive power, he stumbled upon a Lilliputian universe—a realm of tiny life-forms that had previously been imperceptible.

Van Leeuwenhoek was the first person to witness the microbial world—the ubiquitous but imperceptible protozoa, fungi, and bacteria that live among us—and to his eyes they appeared similar to tiny citizens moving about in a miniaturized city square. He called them "animalcules" (microscopic animals) to convey his sense that they were simply miniature versions of the animal world, with senses and internal organs as rich and complex as their larger counterparts. As van Leeuwenhoek's microscopes became more and more powerful, the organisms he detected became smaller and smaller, and there appeared to be no lower limit to how compact living creatures might be.

One would think that microscopes—which could be used to peer inside eggs and sperm, looking for *tiny tiny* bodies—would have put preformationism to rest when such forms could not be found. But to a French priest named Nicolas Malebranche, van Leeuwenhoek's revelations breathed new life into the theory. As Malebranche saw it, the take-home lesson from van Leeuwenhoek's discovery was that our senses deceive us. If an entire community of living creatures could exist in our midst, invisible before the microscope allowed them to be seen, then certainly there must be other invisible worlds waiting to be discovered. Malebranche contended that as more powerful instruments were developed, making it possible to look deeper, microscopists would find evidence of fully intact, preformed bodies inside of eggs.

Indeed, Malebranche's notion of preformation did not stop with a single egg. Rather, he imagined that each preformed body residing in an egg must carry its own eggs, each with their own preformed bodies, and so on, and so on. In other words, he imagined a world in which each egg contained an army of preformed bodies, an infinite succession of "seeds within seeds" like a nesting Russian Matryoshka doll. If van Leeuwenhoek's animalcules had evaded detection by human beings up until that point, who could say whether other miraculous arrangements would be revealed with time?

Malebranche's ideas were rooted in philosophy and theology, but his theories stimulated a reexamination of Aristotle's methods and conclusions. Surprisingly, these investigations also fueled the reemergence of preformationism. The most influential evidence came from Jan Swammerdam, a seventeenth-century Dutch naturalist who was renowned for his fine dissections of insects, which he would occasionally perform at private gatherings for his wealthy patrons. Swammerdam was particularly interested in immature forms—grubs, caterpillars, and chrysalises—and made a remarkable observation when he dismembered them: these larvae and pupae seemed to contain most if not all the organs of the future insect. Legs, wings, abdominal segments, and antennae were already in place before the moth or butterfly was born, crumpled up as if waiting for some signal to unfold.

Even though Swammerdam's dissections didn't show how early in development a body might take shape, his observations offered further support for the notion of ready-made forms. Other endorsements followed, and with support from some of the leading naturalists of the eighteenth century—Marcello Malpighi, Lazzaro Spallanzani, Charles Bonnet, and Albrecht von Haller—preformationism again became the dominant explanation for embryonic origins.

Furthermore, one of van Leeuwenhoek's other observations gave the theory a new twist. The Dutch lens grinder had noted that human sperm had tails and could swim like tadpoles, an indication that they possessed a life force and, perhaps, even a soul. This raised the possibility that sperm, and not the more sedentary eggs, might be the source of the "pre-embryo." This discovery fractured the preformationist movement into two camps—the "ovists," who believed the presumptive body was carried in the female egg, and the "spermists," who believed it was carried by the male sperm. Debates between spermists and ovists took on a life of their own, becoming so loud that they drowned out any question of whether the theory of preformation itself was valid.

Amid these arguments, the clergy and gentry also voiced their support for the theory of preformed bodies. For the church, the phi-

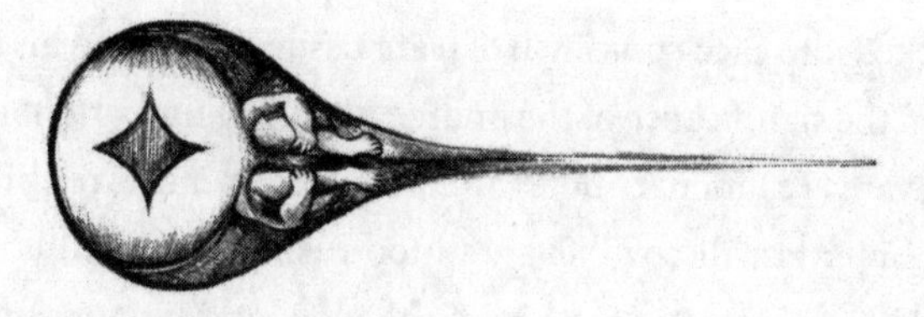

Nicolaas Hartsoeker, a seventeenth-century Dutch microscopist and confirmed spermist, believed that preformed bodies existed within the sperm. Hartsoeker did not actually claim to have observed this directly; his diagram was merely speculative. (REDRAWN FROM HARTSOEKER, *ESSAI DE DIOPTRIQUE*, 1694.)

losophy of a never-ending chain of antecedent embryos served as a link between the earthly and the divine. If Malebranche's claim of infinite seeds within seeds was true, it implied that every creature past, present, and future was fashioned at the moment of creation by a divine spark. It was living proof that God had implanted all of humanity in the ovary of Eve (or, if you were a spermist, in the testes of Adam). For the aristocracy, preformationism legitimized ancestral rights, since it defined as the natural order of things the conviction that every king belongs to a preexisting succession of kings, and every peasant belongs to a preexisting succession of peasants. Preformationism—with its religious, scientific, and social implications—held something appealing for anyone in a position of power, and it persisted as the dominant model of embryonic development for nearly two centuries.

A JIGSAW PUZZLE

THEN, IN THE MID-NINETEENTH CENTURY, TWO THEORIES TILTED THE scales back in Aristotle's favor, or at least balanced them out. The first was *cell theory*—the principle that the cell is the basic unit of life, akin to the atom in chemistry or the photon in physics. The word "cell" was coined in 1665 by the English naturalist Robert Hooke when he noticed that cork, when placed under a powerful magnifying glass, was divided into subunits that resembled the cells of a honeycomb, or monks' cells in a monastery. Hooke estimated that there were more than a billion of these units packed into each cubic inch of plant tissue. But he and his successors dismissed the significance of the finding, thinking the pattern interesting but irrelevant (a cautionary note to all would-be scientists reading this that the most important discoveries are all-too-easily overlooked as mundane).

In 1839, botanist Matthias Schleiden and anatomist Theodor Schwann, both German, revived Hooke's discovery, proclaiming that the cell was the irreducible unit of biology and that nothing smaller than a cell could be considered alive. Plants and animals, they declared, were composites of cells, a separate class of life from van Leeuwenhoek's single-celled animalcules. It was, in retrospect, stating the obvious. For two centuries since van Leeuwenhoek's discoveries, biologists had been looking at cells without recognizing their part in the greater whole, seeing the forest but missing the trees. A slew of new questions followed: What do cells do? How do they work? How do old cells give rise to new ones? By the 1850s, this way of viewing the world of biology through cells was prompting a reexamination of every assumption that had come before.

The other game-changing idea was *natural selection*, Charles Darwin's insight that the rich diversity of animals on earth results from chance, not design. Even before Darwin, biologists had come to accept the principle of evolution—the notion that new species emerged gradually over time—but they quarreled over the details. Transmutation, the popular theory championed by French naturalist Jean-Baptiste Lamarck, posited that evolution is the consequence of use or disuse, causing a body part either to grow or to atrophy. As evidence, Lamarck famously pointed to the giraffe, claiming that the creature owed its long legs and neck to previous generations stretching to reach the heights of the acacia tree in the African savanna. In Lamarck's world, new species arose as a consequence of need.

Darwin's radical idea was that competition, rather than adaptation, drives the formation of new species. He argued that nature constantly produces new variants—chance deviations that alter the size, shape, or function of some body part. Most of the time, he reasoned, these variants amount to mistakes, resulting in offspring with a reduced capacity to reproduce. But occasionally such deviations might yield offspring whose chances of survival and procreation—their relative "fitness"—outstrip those of their peers, giving their progeny a selective advantage. In Darwin's world, where the fit are selected and the unfit die off, new species arose independent of need.

As these two great ideas—cell theory and natural selection—percolated through the lecture halls and scientific societies of the nineteenth century, an obvious question presented itself: How does evolution work at the cellular level? Because all traits—including those responsible for evolution—must pass from one generation to the next through the fertilized egg, understanding the details of this hereditary handoff became a priority.

But an obstacle stood in the way: there was no precedent for asking this kind of question. Until that moment, biology had been dominated by naturalism, the tradition of making observations in the field and building theories around them through deductive reasoning. The approach spawned many important ideas (natural selection being perhaps its greatest triumph), but it offered few provisions for real-world validation. For every naturalist theory that has stood the test of time, dozens ultimately collapsed. (One of these, as we will see later, was Darwin's own flawed theory of heredity.)

By the turn of the twentieth century, the tradition of naturalism was giving way to a fresh appreciation of the animal world based on experimentation. To better understand the dichotomy between naturalism and this new approach, which we can call experimental biology, consider the following analogy. Imagine that you have been given a machine (for example, a pendulum clock) and asked to figure out how it works. The naturalist approach—observation and deduction—could teach you quite a bit. You might note how the hands move with a regular periodicity and how each hand's rotation is related to that of the other two. If you looked carefully enough, you might notice that the periods of the second hand, the minute hand, and the hour hand are related to each other by a factor of 1:60. But if you were asked what makes the clock go—what causes the hands to move or how they are connected to each other physically—observation alone would leave you guessing. The only way to answer this question, to gain an understanding of how the thing works, would be to open it up, look inside, and start tinkering with the parts until the mechanisms became clear.

Instead of merely inspecting embryos, scientists began to manipulate them, asking how disruption of one part affected the whole. The naturalists, having spent their careers in the field creating ever-more-detailed taxonomies, were dubious that anything important could be learned by studying cells in the isolated environment of a laboratory. But like any disruptive technology, experimental biology was about to transform the world of science.

EMBRYOGENESIS BEGINS WITH FERTILIZATION, THE FUSION OF TWO *gametes*—the "half cells" of sperm and egg—into a single-celled embryo, or *zygote*. Even before Aristotle's time, it was understood that everything needed to form a new animal—a chicken, for example—could fit inside a fertilized egg. But cell theory made it clear that bodies grow through the accumulation of new cells, implying that cells must divide. By the late 1800s, microscopists had developed methods to observe the phenomenon directly, showing that when a cell splits into two cells, small threads within it (later called *chromosomes*) were pulled apart and deposited within the resulting daughters.

The cell divisions that occur early in embryogenesis—when the zygote expands to become a cluster of several hundred cells—are unlike those during any other stage of life. Under most circumstances, division is preceded by a period of cellular enlargement, so that the size of each daughter cell (after division) is roughly the same as that of its parent. But in the early hours of embryogenesis, there is no growth before division. Instead, a thin wall forms in the middle of each cell, splitting it in half and resulting in two new, smaller cells. The process repeats itself over and over, reducing each cell's size twofold in a process known as *cleavage* until the embryo is a spherical ball of cells known as a *blastula*.

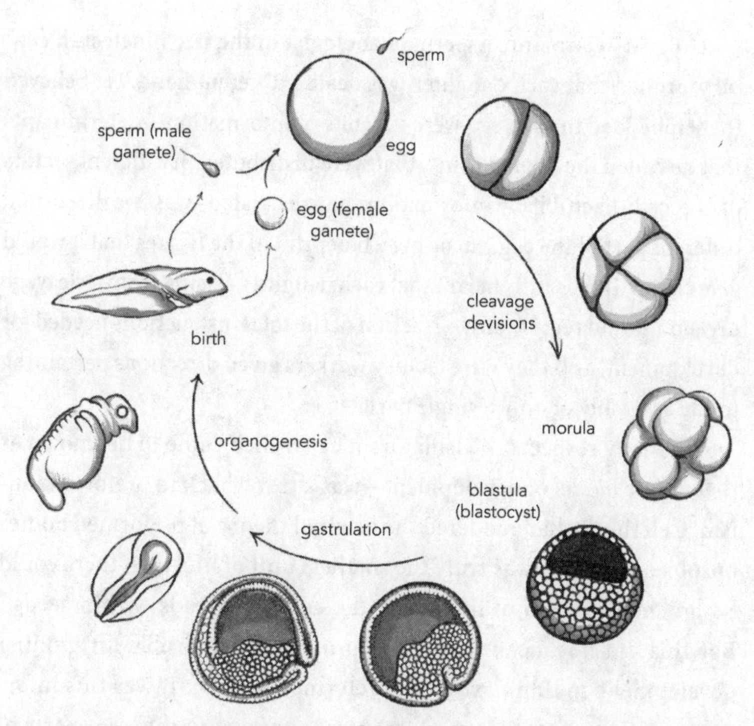

Following fertilization, the single-celled embryo undergoes cleavage, its cells shrinking with each division. After several divisions, the embryo is considered a morula, and once it has more than 100 cells it is called a blastula (or blastocyst in mammals). As the embryo now grows bigger, gastrulation results in three germ layers from which all organs form (organogenesis). The embryo continues to mature until it is ready to be born, complete with its own gametes—egg or sperm—that can repeat the cycle.

Collectively, these observations prompted a series of questions about the nature of development and the instructions contained within the zygote: What happens to those instructions when a cell undergoes division? Are developmental directives partitioned with each cleavage, apportioned to the two daughter cells (*blastomeres*) the way each player in a game of poker is dealt a different hand from a deck of cards? Or does each cell get an identical and full set of instructions, as if given a complete deck? Lastly, and most intriguingly, where, and in what form, is the information for development stored?

August Weismann, a German zoologist in the late nineteenth century, argued that each daughter was dealt a different hand. He believed that embedded in each egg were tiny bits of information—determinants that he called the *germ plasm*—that were distributed in different sectors of the cell. Each bit of information, he speculated, was localized to a different part of the egg, encoding a blueprint of the tissues that it would give rise to. If true, this meant that each daughter arising from a cleavage division would receive only a fraction of the total instructions needed for development, as if they were factory workers given directions pertaining to the assembly of only a single part.

In many respects, Weismann's idea—which came to be known as the *mosaic model* of development—was a throwback to preformationism. Cell theory had rendered the original theory of preformed bodies obsolete: if the cell was truly the smallest unit of life, then there could be no smaller form of life inside the egg—no "seeds within seeds." But this did not mean that the *information* responsible for guiding development couldn't exist in a preformed pattern. It was this information, Weismann theorized, that was confined to different parts of the egg. As he saw it, the zygote was like a presolved jigsaw puzzle, the destiny of each piece determined by the part of the egg it derived from. As biologist Stephen Jay Gould later put it, "If the egg were truly unorganized, homogenous material without preformed parts, then how could it yield such wondrous complexity without a mysterious directing force?"

HALF EMBRYOS

THE MOSAIC MODEL HAD OTHER THINGS GOING FOR IT AS WELL. IN the late 1800s, biologists had discovered "organelles," tiny organ-like subunits within cells whose purpose was unknown. The *nucleus*, which later proved to be the seat of the cell's genetic material, was one such compartment, but others were being described, each with a distinct shape: mitochondria, golgi, networks of endoplasmic reticulum, and so on. It was easy to imagine how one or more of these fragments, distributed among daughter cells in an uneven manner, could drive cells to develop along independent paths.

In 1888, German physician Wilhelm Roux (pronounced "Roo") put Weismann's mosaic model to the test. One of the leaders in the revolt against naturalism, Roux believed that observation alone could never provide a satisfying picture of biology. Rather than theorizing about the behaviors of tissues and cells, as the naturalists had done, he sought to understand the full chain of events underlying the development of an animal, something best achieved through experimentation. Only in the controlled environment of the laboratory, Roux insisted, could one pry a biological specimen apart and manipulate the pieces, as one would do to understand the workings of any mechanical device. "Speculation must be checked against reality" was Roux's credo.

A second pillar of Roux's philosophy was that animals do not possess unique qualities—no spark, soul, or spirit—setting them apart from inanimate objects. He considered everything in the natural world to be a product of physics and chemistry, subject to their fundamental laws. Frogs and humans, he believed, are sculpted by the same forces that shape mountains and streams; the two categories differed only in the degree of complexity. But when it came to the One Cell Problem, Roux was thrown for a loop. It was hard for him to conceive of how the billions of cells that make up an animal come into being, travel to the right place, and perform the right function—all starting from a single cell. Most confusing of all: How could the laws of physics and chemistry explain an entity that seemed capable of assembling itself?

Weismann's model offered an attractive answer. If eggs came preassembled—each domain destined to give rise to a designated portion of the future animal—this could help explain the embryo's miraculous abilities. Roux imagined bits of information squirreled away in different parts of the egg, a spatial choreography that guided the embryo through its developmental dance. But the mosaic model was only that—a model—and Roux needed proof.

In the end, he settled on a strategy of "cellular ablation." The approach involved killing a blastomere at the earliest stages of embryonic development and seeing what happened to the survivors. Roux reasoned that if Weismann's theory was correct, then the fertilized egg should pass on to each daughter only a portion of its developmental instructions when it divided. Consequently, if one of these two daughter cells were to die, the survivor (lacking a full set of instructions) would be expected to give rise to only a portion of the embryo. If, conversely, the mosaic model was wrong, then the surviving cell might develop as if nothing had happened, ignoring the unfortunate fate of its sibling.

For these experiments, Roux chose to use frog eggs, which are large and can be fertilized in the controlled environment of the laboratory. Watching intently through the eyepiece of his microscope, he waited for the zygotes to begin their first cleavage, a process heralded by the formation of small furrows on either side of the freshly fertilized eggs. Once the separation was complete, he used a red-hot needle—its tip much smaller than a cell—to dispatch one of the two daughter cells. Then, he moved on to the next embryo, and the next, a serial killer of blastomeres intent on seeing the survivors' response.

When he inspected the traumatized embryos the next day, the results fulfilled the predictions of Weismann's theory. While each cellular corpse remained an amorphous blob, its twin continued to mature, carrying out its normal developmental program without apparent regard for its surroundings. The result was a petri dish filled with bizarre "hemi-embryos," in which one side of the embryo developed normally and the other remained lifeless. Roux took the experiment a step further, waiting for the embryo to cleave a second time so that four blastomeres were present before he applied his needle. The results were consistent with his first set of experiments: if he killed one of the four blastomeres and allowed the remaining three to develop for several hours, the result was a "three-quarter" embryo missing a fourth of its body. Conversely, killing three of the four blastomeres at the same stage and letting the resulting cell grow resulted in a "one-quarter" embryo. Whatever he

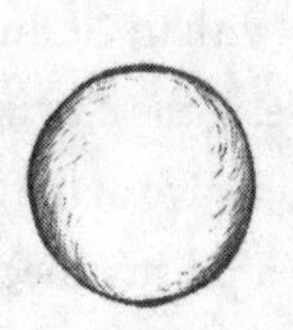

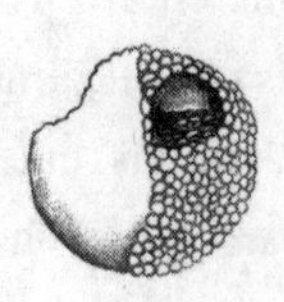

Wilhelm Roux believed that embryonic cells followed an autonomous, or mosaic, developmental path. To test this, he used a hot needle to kill one of the blastomeres in a two-celled frog embryo. The surviving cell formed only half an embryo, which Roux took as evidence that each cell developed in an autonomous manner.

did, the remaining cells carried on autonomously, seemingly oblivious to their neighbors' fate.

At face value, this was strong evidence for the mosaic model. How else could one explain the independent trajectory of the surviving cells if they weren't "preprogrammed" in some way? Roux was not alone in this interpretation. At around the same time, French biologist Laurent Chabry was conducting experiments on the sea squirt, a potato-shaped *invertebrate* that adheres to ship hulls and to rocks on the sea floor. Chabry found that, if separated, sea squirt blastomeres followed their prescribed developmental paths, another indication that each made a unique contribution to the whole. Like Weismann, Roux concluded that the blueprints for development must be distributed in the egg in a spatially organized manner, imperceptibly hardwired into its three-dimensional fabric. But in reaching this conclusion, Roux had made a fatal miscalculation, one that came to light—like so many discoveries in science—by accident.

A MACHINE THAT COULD REBUILD ITSELF

THE AGENT OF THAT SERENDIPITY WAS HANS DRIESCH, A 22-YEAR-OLD scientist who had only recently earned his PhD. Driesch yearned to see the world, and after completing his dissertation in 1889 he traveled to the Far East, where he absorbed philosophies that saw the natural world in holistic terms different from those he had grown up with in Germany. En route back to his native country, he paused in Naples—the city in the shadow of Mount Vesuvius that Alexandre Dumas referred to at the time as "the flower of paradise." What was intended to be a pit stop became Driesch's home for the next decade.

Beyond being one of the most vibrant European cities of the late nineteenth century, Naples was also home to a recently built center for biomedical research called the Stazione Zoologica. The Station, as it was nicknamed, was establishing a new paradigm for research in which scientists could rent a laboratory bench and equipment, just as an artist might lease studio space and supplies. The model was hugely successful, and scientists flocked there from around the world, eager to pursue their research free of the competing demands of university life (for those lucky enough to be granted a leave of absence). Moreover, the location could not have been better—the Station was situated a stone's throw away from the Bay of Naples, giving researchers easy access to the marine creatures most of them used in their studies.

Driesch had grown up in a wealthy household, and this relieved him of the need to take on a university post—a freedom that undoubtedly irritated his coworkers at the Station, who all had cumbersome teaching responsibilities waiting for them back home. While his primary reason for remaining in Italy had been the Station's growing scientific reputation, the Neopolitan nightlife provided a delightful backdrop. He took full advantage of his bachelor's existence, using Naples as a jumping-off point for trips throughout the Mediterranean, northern Africa, and Asia.

In all these respects, Driesch was entirely different from Roux, whose serious temperament had little appetite for the pleasures in which Driesch indulged. When it came to science, the two men also differed significantly. Roux was meticulous and adroit at the bench, focused on every detail. If the scientific question called for it, Roux eagerly built a new device or invented a new technique. And when it came to performing the experiments, Roux's ability to carry out the most delicate operation was unparalleled—a feature known, in the vernacular of the laboratory, as "good hands."

Driesch, by contrast, did not have good hands. He was clumsy, and he lacked Roux's patience and dexterity. Consequently, he sought out shortcuts—ways of doing an experiment that did not require sophisticated gadgets or fine motor skills. Ironically, it would be this parsimony that allowed Driesch to see what Roux had missed.

LIKE MOST OF HIS CONTEMPORARIES, DRIESCH WAS EXCITED BY Roux's supposed confirmation of the mosaic model, and he was anxious to repeat the experiment as a way for him to get acclimated to the Station. But lacking the ability (and the inclination) to use Roux's painstaking protocol, he sought a simpler technique. Instead of frogs, Driesch decided to use sea urchins, spiny palm-sized animals with thousands of antenna-like sensory organs. Like frogs, sea urchins have large eggs, which would make their early embryos easy to study. Another advantage was that they lived in the bay, which would simplify collecting them. But Driesch's overriding reason for choosing this marine organism was its resilience. "The sea urchin egg was able to put up with my interventions without dying," he later confessed.

Another deviation from Roux's methods involved a change in tactics. Instead of killing cells, which required great dexterity, Driesch would simply separate them, splitting the embryo like a modern-day Solomon. Friends of his at the Station, brothers Richard and Oskar Hertwig, had found that shaking sea urchin embryos vigorously could cause their blastomeres to disconnect. In principle, this would allow Driesch to study mosaic development using a different and simpler method, the only difference being that, where Roux had followed the fate of blastomeres whose sisters had been killed, Driesch would follow the fate of blastomeres that had been isolated from each other.

In the summer of 1891, he proceeded with the experiment. After collecting dozens of sea urchins, he isolated their eggs and sperm and mixed the two together in a test tube to allow fertilization to occur. With predictable timing, the zygotes completed their first cleavage division, filling the tube with two-celled embryos. This was Driesch's cue. He shook the tube vigorously for a few moments, causing the blastomeres to disperse, leaving only single cells where previously there had been doublets. Over the next several hours, the manhandled blastomeres continued their metronomic divisions until each resembled a grape-like cluster of cells—a blastula. Then, with a much softer touch, he transferred the embryos to petri dishes, filled them with fresh seawater, and left the embryos to mature overnight.

Driesch expected the morning to bring with it some malformed but (hopefully) recognizable bits of tissue, the sea urchin equivalents of Roux's half embryos. Instead, the dishes were alive with movement. Normal-appearing sea urchin larvae, smaller than average but unremarkable in all other respects, were busily swimming around the dishes. Each blastomere he had shaken loose seemed capable of near-normal development—an almost unbelievable result. He repeated the experiment several times, and the outcome was always the same. Then, following Roux's footsteps, he waited until the fertilized sea urchin egg had undergone a second cleavage division before shaking. And again, remarkably, each blastomere generated a new animal on its own, a far cry from the "one-quarter" and "three-quarter" embryos Roux had observed in his frog experiments.

When word of Driesch's results eventually reached Roux, the venerable professor refused to believe it. (It probably didn't help that Driesch unequivocally stated that Roux's theory now had to be discarded.) In Roux's hands, *cell fate* had a determined or autonomous quality, seemingly unconcerned with what was happening in its surroundings. In Driesch's hands, cell fate was conditional, dependent on whether it was alone or tethered to its sibling. The results were incompatible. Was the discrepancy the result of a difference between frogs and sea urchins? Had someone made a mistake? Could both results be right?

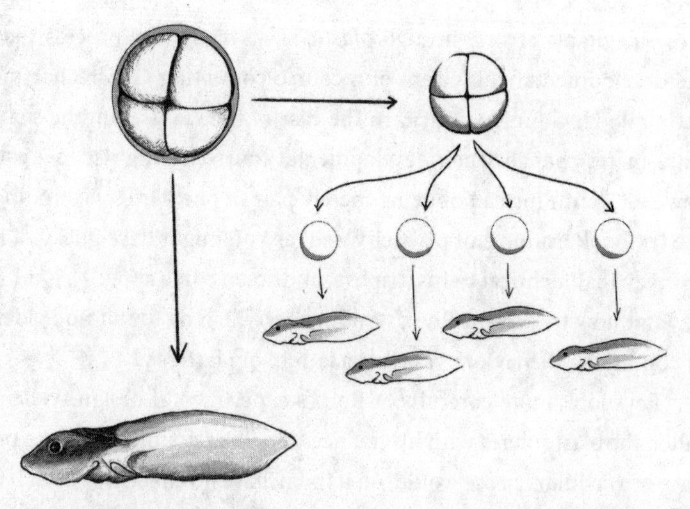

Hans Driesch separated blastomeres instead of killing them and found that each blastomere was able to give rise to a complete embryo. In contrast to Roux's findings, this result showed that the fate of each cell in the early embryo is conditional–able to change based on its surroundings.

The two researchers had approached the embryo as one might approach a complex machine, seeking to understand its workings as if it were a clock or a car. But a machine that could rebuild itself after being halved or quartered—how does a person wrap their head around that?

THE JOURNEY OF DEVELOPMENT IS UNPREDICTABLE, A VOYAGE fraught with uncertainty and peril. Each time a cell divides, it reads, copies, and interprets billions of letters of DNA, repeating the procedure countless times before an animal reaches adulthood. Most molecular processes in the cell are remarkably precise, with an accuracy often exceeding 99.9 percent. But the sheer scale of the enterprise—a venture that can involve trillions of cells—means that gaffes are inevitable. (If a cell must perform a billion operations each time it divides, then even with an error rate of 0.01 percent, thousands of mistakes will occur.) Nature copes with these inevitable errors through *plasticity*—a regulatory process that is the developmental equivalent of a course-correcting GPS recharting a path following a missed turn. In the case of the sea urchin, the shaken blastomeres changed their developmental course, giving rise to a whole new sea urchin instead of being merely part of one. Driesch's finding is the textbook example of plasticity and, as you might have guessed, it is one way that identical twins, triplets, and so on come about.

But how to explain Roux's finding, in which he found no evidence of conditional behavior? Was this a failure of plasticity?

Let's look more carefully at Roux's experimental design. When he killed the blastomeres with his hot needles, Roux assumed that the presence of a cellular corpse would, on its own, have no effect. But what if this assumption was false? Could the dead cells, lacking any developmental potential of their own, still influence their siblings? The answer, improbably enough, turned out to be yes. Even after Roux's lethal intervention, the cellular remains, although lifeless, were still capable of sending messages from beyond the grave. And the message those cells delivered was: "I am still here!" The survivors of Roux's interventions, believing their siblings still present, and deferring to the signal from their phantom other halves, curtailed their own development to form hemi-embryos instead of complete ones.

Without realizing it, Driesch had sidestepped the issue with his shaking; dispersed and removed from the inhibitory influences of their neighbors, the sea urchin blastomeres could access their internal GPSs and recover the potential they previously possessed as zygotes.

When Roux and Driesch did their experiments, such notions of cellular plasticity were counterintuitive. How could a cell seemingly, and suddenly, "change its mind"? Driesch, the idea's architect, found the concept more troubling than anyone else. Embryogenesis was far too accurate, too reproducible, to imagine something as important as a cell's fate being left to chance. Unable to comprehend how a cell might adapt to changing circumstances, Driesch reached for alternate explanations. He turned to *entelechy*, a term originally coined by Aristotle meaning "vital force" or "soul," as the explanation for his inscrutable observations. How else could a cell change its behavior so dramatically, if not for some guiding mystical influence? By 1910, he had abandoned experimental biology altogether, pursuing philosophy, parapsychology, and even psychic research for the remainder of his career. Although he had poked holes in Weismann's mosaic model, he lacked a new model with which to replace it. To many, it seemed that Driesch had lost his mind.

THE ORGANIZER

IN 1896, A 27-YEAR-OLD GERMAN BIOLOGIST NAMED HANS SPEMANN came down with a case of tuberculosis. The illness struck just as he was completing his studies, which spanned medicine, zoology, and physics. To combat the boredom of a long convalescence, the bed-bound biologist asked his caretakers to deliver a constant supply of scientific reading material. Devouring anything he could get his hands on, Spemann happened upon a volume with the pithy title *Das Keimplasma: Eine Theorie der Vererbung* ("The Germ-Plasm: A Theory of Heredity")—it was Weismann's treatise, the discourse that outlined the great biologist's ideas on heredity and development. Spemann had not given much thought to embryology or the One Cell Problem beforehand, but now he was hooked. After reading all of Weismann's works, he moved on to Roux, then Driesch, then the commentary on their work.

The material left him with the same question that had driven Driesch away from experimental science—how does one study a machine (the embryo) that can seemingly reinvent itself? Was there perhaps another method for studying the "decisions" made by embryonic cells besides ablating them or separating them from their neighbors? The question nagged at him, and from his sickbed he imagined the various ways that, once he was better, he might fill this void.

His brainchild was an experimental approach called *cellular transplantation*—a method that involved moving cells from one embryo to another. During the transplant procedure, the uprooted cells could be placed anywhere on the recipient's body—either the same general area from which they had originated in the host, or an entirely new region. Spemann reasoned that, by observing what those cells did following their engraftment, he could determine whether their destiny was fixed (deterministic) or conditional (plastic). It was the experimental equivalent of the George Bernard Shaw play *Pygmalion*, in which a poor flower girl, Eliza Doolittle, is plucked from the streets of London and relocated to the high-society home of linguist Henry Higgins. The question at hand in the play was whether she would fit in with her new surroundings (she did). As a biologist, Spemann saw transplantation in a similar light: it was a way to determine whether embryonic cells were preconditioned to follow a given developmental path or whether they could adapt to new environments, the property of plasticity.

After his recovery, Spemann began putting these ideas to work using embryos from the common striped newt. Like Roux, Spemann was blessed with good hands, a knack for working with fragile specimens. This turned out to be a necessity, as cellular transplantation required fine motor skills of the highest order. Spemann developed his own tools for holding and moving embryos—hair loops for cutting and extracting tissue, and beaded micropipettes for pushing the grafts into place. All this had to be done while staring through a microscope eyepiece and adhering to strict sterile conditions, as bacterial contamination could easily ruin a day's work.

As Spemann perfected his technique over the next two decades, moving hundreds of patches of cells from one embryo to another, a consistent pattern emerged: the relocated cells almost always took on the attributes of their new environment. If he moved cells from the host's back to the recipient's belly, the transplanted cells took on the identity of "belly" as if they had belonged there all along. Other variations followed the same pattern. Regardless of the source of the tissue, or its destination, the relocated cells readily adapted to their new surroundings. It was like seeing a microscopic Eliza Doolittle assimilating into London high society—an unmistakable illustration of plasticity.

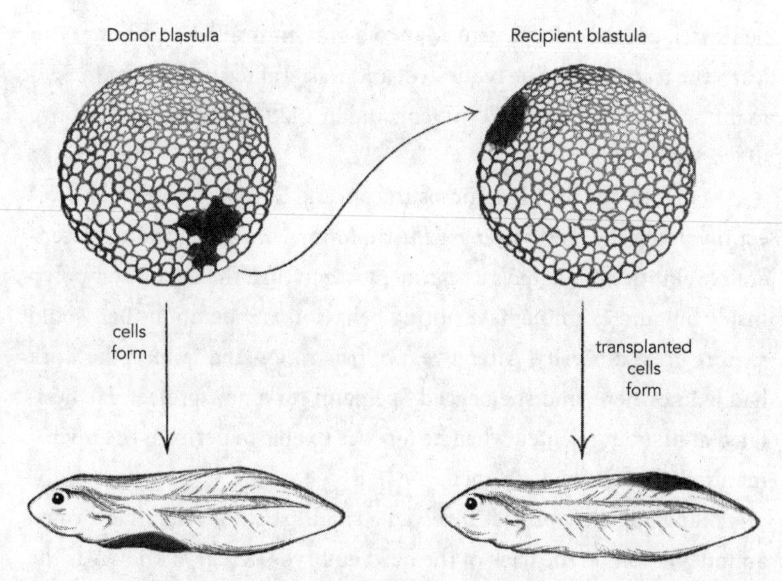

Spemann's cellular transplantation experiments in young embryos was a further demonstration of the principle of cell plasticity.

Spemann also discovered a second, important feature of this trait: a graft's plasticity depended on the donor's age. As long as the source of the transplanted tissue was still young (that is, a blastula), the relocated cells acclimated to their new environments. But if the transplants were performed later, once the embryo had started to mature, the cells tended to behave as they would have in the host. In other words, cells became more *committed* to their developmental path as the embryo aged.

Plasticity, it turns out, is a privilege of youth.

THE STORY MIGHT HAVE ENDED THERE WERE IT NOT FOR HILDE MANgold, a 22-year-old student at the University of Frankfurt. By chance, Mangold attended a lecture given by Spemann, who was in town as a visiting professor. Spemann made no secret of the technical difficulties with cellular transplantation. But rather than being put off by the challenges he described, the effervescent Mangold was enthralled and hungry to learn the techniques. She requested, and was granted, a position in Spemann's laboratory as a PhD student, and in 1920 she moved to Freiburg to work with him.

Mangold was a skilled microsurgeon, and Spemann gave her a project involving the freshwater *Hydra*, an animal with astounding regenerative abilities. Mangold's assignment was to turn the freshwater polyp inside out and examine its resulting behavior, an operation that would require great dexterity. After a year of frustration and failure, the work had led nowhere, and she begged Spemann for a new project. He hesitated at first but relented when he too was unable to perform the micromanipulations he had assigned.

Mangold's new project involved a small patch of cells that formed an indentation on the back of the newt embryo—a region known as the dorsal lip. Spemann's earlier studies had shown that the dorsal lip was the point of initiation for *gastrulation*, a critical process during which the embryo undergoes dramatic changes. We will have more to say about gastrulation later, but for now, think of it as a series of cellular acrobatics by which the dorsal lip grows from a small depression into a large sinkhole, eventually swallowing up the surrounding cells and carrying them to the embryo's interior. Spemann had grown interested in this group of cells because they seemed to defy the rule of plasticity in youth. Unlike other cells in the blastula, which adapted to their new environment when transplanted, dorsal lip cells behaved just as they would have if left in place, creating a cavity in preparation for gastrulation. Clearly, there was something unique about this tiny patch of tissue, and Mangold was charged with investigating further.

To fully understand what was happening, Mangold needed to track the behavior of both the transplanted cells and their neighbors, which required a means of distinguishing donor from host. The method for doing so came from Spemann himself, whose earlier work had relied on two related newt species—one with darkly pigmented cells (*Triton taeniatus*) and one with faintly pigmented cells (*Triton cristatus*). Because the two species were closely related, they could accept embryonic grafts from each other, but the pigmentation differences made it possible to unambiguously determine each cell's origin, whether donor or recipient.

In the spring of 1921, Mangold performed hundreds of transplants between the two kinds of newt embryos. Bacterial contamination overwhelmed most of her experimental subjects, leaving only a single graft that survived to maturity. And yet one was enough: amazingly, a second embryo had started to grow at the graft site, a kind of conjoined twin. Another surprise came when Mangold inspected the pigmentation of the new twin. Although she expected to find that the graft itself was giving rise to this ectopic outgrowth, she observed instead that the new embryo contained cells from *both* donor and host. This could mean only one thing: the transplanted cells had coaxed their new neighbors to adopt a different developmental course, one that led to the formation of a mirror-image embryo. This fate-altering conversion became known as embryonic *induction*, and it was the cellular equivalent of Eliza Doolittle forcing a lord to become a Cockney.

There was no longer any doubt: cells "talk" to one another during

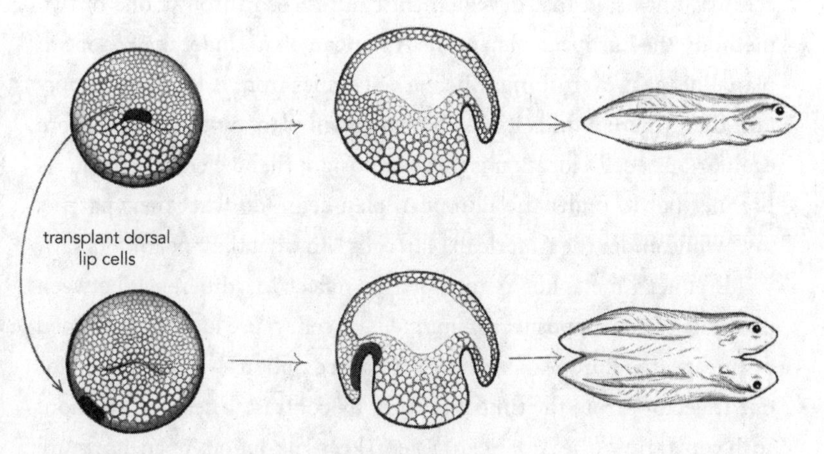

Spemann and his student, Hilde Mangold, found that the dorsal lip cells of a gastrulation-stage embryo had a special property. This tissue, called the organizer, could induce the formation of a second embryo.

development, using a form of communication that only they could understand. Driesch's results had hinted at the existence of induction, but now scientists could study the phenomenon directly. In sea urchins and frogs, messages passed between neighboring cells could halt the generation of a sibling. In the newt blastula, messages from a small patch of cells had the opposite effect, prompting the formation of a twin.

To capture the power held by the cells of the dorsal lip, Spemann gave the little patch of tissue the impressive-sounding name the *organizer*. Its existence solidified induction as a central concept in embryology, earning Spemann a Nobel Prize in 1935. Mangold, Spemann's student who had conducted the crucial experiment, never got to see the impact of her discovery, nor partake in the Nobel Prize. The 26-year-old mother died tragically when a gasoline heater in her kitchen exploded in September 1924. Her paper had just come out.

THE AMERICAN SYSTEM AND THE EUROPEAN SYSTEM

SYDNEY BRENNER, ONE OF THE GREAT BIOLOGISTS OF THE TWENTIETH century, once said that development can proceed through one of two methods: the European plan or the American plan. Under the European plan, lineage is everything: where a cell comes from is far more important than where it finds itself. The American plan, by contrast, is more egalitarian: a cell's location is more important than where it is from. As Brenner put it, under the European plan cells "do what their parents say," while under the American plan cells "do what their neighbors say."

Brenner's metaphor is intended to reflect the difference between plasticity and its opposite, *commitment*. A cell whose identity is defined by lineage (the European system) has limited options—it is groomed for one trajectory from the time it is born. By contrast, a cell born without such constraints (the American system) keeps its options open; its future path will be determined by its experience. Embryos are a mix of both plastic and committed cells, with the balance tilting toward plasticity early in development and commitment later. Cells become set in their ways as they age, just as many people do.

Viewed differently, Brenner's metaphor is a way of restating the classic debate between nature and nurture: the argument over whether our biology is predetermined and "hardwired" or indeterminate and open to external influence. The issues we have encountered so far—the dispute between the preformationists and the epigenesists, played out in the more modern distinction between commitment and plasticity—all speak to this tension between nature and nurture. But as experience and common sense tell us, neither holds a monopoly.

There are times when the American system dominates and times when the European system dominates. Plasticity provides embryonic cells with the potential to change course and adapt to the inevitable slip-ups inherent to development. But if every cell's fate depended solely on the behavior of other cells—if everything were relative—there would be nothing to give the embryo its structure in the first place. Commitment—the loss or absence of plasticity—is the source of that structure, providing developing cells with stable frames of reference, landmarks they can use to orient themselves. For development to proceed normally, some cells must remain open to change and others must pledge themselves to a specific vocation—a compromise between adaptability and edict. Our embryonic origins are thus laced with paradox: the single cell that gives rise to an animal bears an innate trajectory yet retains the capacity to change that trajectory. In this respect, Driesch was right. The embryo is unlike any other machine.

It is this paradox that prompted past generations to imagine animals existing in a preformed state inside the egg, like the "tiny tiny babies" my daughter pictured. Neither Roux nor Driesch could wrap their heads around the conflict between the fixed and the flexible, and Spemann, who witnessed the dichotomy between nature and nurture almost daily, could only guess at its underpinnings. There is indeed an intangible developmental force pushing the embryo—but it is different from entelechy or the "soul" that Driesch imagined. Rather, the entity that gives development its momentum, and referees the balance between commitment and plasticity, is subject to all the laws of chemistry and physics. For the mysterious external force that Driesch perceived—as did Aristotle millennia earlier—is the invisible hand of evolution, acting over billions of years to fashion the developmental blueprint that each zygote carries.

The solution to the One Cell Problem, it turns out, is embedded in our genes.

Chapter 2

THE LANGUAGE OF THE CELL

On what slender threads do life and fortune hang.

— ALEXANDRE DUMAS, *THE COUNT OF MONTE-CRISTO*

Time flies like an arrow. Fruit flies like a banana.

— ATTRIBUTED TO GROUCHO MARX

As laboratories go, room 613 in Columbia University's Schermerhorn Hall—or the Fly Room, as it was more affectionately known—was cramped. At only 16 by 23 feet, the size of a spacious living room, the laboratory barely accommodated the eight desks that were crowded together inside. Bits of decomposing fruit and other detritus—leftovers from the diet for fruit flies that gave the lab its nickname—had made their way into the room's crevices and corners, creating a secondary ecosystem for cockroaches and mold. The lab's human occupants, their names now legendary, were oblivious to the squalid surroundings, too absorbed in their own experiments (or in searching for flaws in those of their colleagues) to take notice. To a visitor from the outside, it was hard to see how any work got done between the grime and the noise (and one can only imagine the smell!). Yet it was here that the worlds of *genetics* and chemistry first met, resulting in a union that would change the course of biology.

The Fly Room was the workshop of Thomas Hunt Morgan, a Kentucky-born embryologist with a keen intellect and a love of collaboration. Morgan had grown up on a scientific diet of naturalism—the eighteenth-and-nineteenth-century approach to science that involved, as Darwin described it, "patiently accumulating and reflecting on all sorts of facts which could possibly have any bearing." As a student, Morgan had mastered the naturalists' methods, and his PhD thesis at Johns Hopkins University—an intricate account of sea spider development—was a model of its descriptive methodology.

In 1890, Morgan took a faculty post at Bryn Mawr College as one of only two biology instructors, saddling him with a teaching load that left little time for experimental work. Still, he kept up with advances in the scientific literature, most of which originated across the Atlantic. Under the influence of Roux and his followers, the Europeans were changing the rules of biological inquiry. Less and less absorbed with simple observation, they were becoming more concerned with mechanism—the cause-and-effect rules by which an organism comes into being. In manuscript after manuscript, Morgan witnessed these researchers turning their backs on naturalism in favor of a more mechanistic, causative approach. He began to question whether the methods to which he had so far devoted himself might be flawed. After four years of teaching undergraduate biology, Morgan could sit on the experimental sidelines no longer. He yearned to see this new approach to science in action, and that meant going where the action was.

Morgan's education in Europe, and his subsequent accomplishments in the Fly Room, would change the trajectory of science, giving physical substance to what had been fuzzy ideas about evolution, heredity, and genes. But before we can understand his journey—and the revelations that followed—we need to go further back in time to understand where our ideas of heredity came from, starting from a period before we knew there were such things as genes or DNA.

A SHAKY FRAMEWORK

THE SCRIPT OF DEVELOPMENT IS WRITTEN IN THE LANGUAGE OF genes. They influence every aspect of life, from the sex, shape, and size of an organism to its health and behavior. They mold us from the moment of conception, not only shaping our physical features but also influencing personality, intelligence, disease, and longevity. Genes impact every step during development, guiding an embryo's fate from the moment of fertilization until birth, and beyond. Whatever befalls us in life—our health and happiness, strengths and weaknesses, successes, challenges, or failings—our genes have something to say about it, even if it is only a whisper. It is a splendid language, rich with idioms and a (literally) evolving vocabulary. Having grasped the basics of alphabet, punctuation, and syntax, molecular biologists have become masters of this genetic tongue. And as we will later see, not only can we read and write in the vernacular of genes, we can also edit it.

Yet a century and a half ago, the existence of such a genetic language could barely be imagined. Our fluency came about gradually through two paths of discovery—that of the geneticists who defined the rules of inheritance, and that of the biochemists who focused on the material basis of heredity—paths that were parallel but separate before merging in the mid-twentieth century. It was only then, with the union of these two fields, that developmental biologists could begin to interrogate the One Cell Problem in molecular terms.

HISTORICALLY, HEREDITY WAS NOT A TOPIC OF DEEP CONTEMPLAtion. The most common framework to explain inheritance was "blending," the idea that children are amalgams of their parents, combined as an artist might mix red and yellow paint to make orange. It was a convenient concept that matched the observations that a child's physical characteristics—its height, facial features, and complexion—typically fall between those of its parents. The concept also resonated in the agrarian world, where farmers since prehistoric times had crossbred plants and animals with desirable features to capture the best attributes of both parents in the offspring.

But in other respects, blending fell short. Eye color was one example. A child born to a blue-eyed mother and brown-eyed father did not develop eyes with an in-between color, as blending would have predicted; instead, their eyes were *either* blue or brown. Hemophilia, the "royal disease" endemic to the noble houses of Europe, was another example. There was no doubt that the disease was inherited, but because only males were afflicted with the tendency to bleed, the trait appeared to "skip" generations. Nevertheless, blending remained the favored explanation despite these exceptions, and few biologists found the subject of heredity stimulating enough to devote serious time and attention to it.

Interest in heredity grew with Darwin's theory of natural selection, the game-changing idea introduced earlier that new species emerge when a fresh trait, arising by chance, makes its recipients more competitive. Because the concept rested entirely on the transmission of such traits from one generation to the next, heredity suddenly became the linchpin of evolutionary theory. Darwin's original treatise offered no explanation for heredity (he, too, believed in blending), and it was a gap that his critics pounced on.

In response, Darwin took a stab at filling the breach in 1868 with an adjunct theory called *pangenesis*. The theory postulated that tissues released tiny hereditary fragments called *gemmules*, which circulated throughout the body before settling in the egg and sperm, where they could be passed along to the next generation. Importantly, Darwin imagined that environmental pressures acting on an animal during its lifetime could modify the makeup of the gemmules, altering their composition in a manner that could be inherited by the next generation, such as a longer beak or a more upright posture. Rather than contradicting Jean-Baptiste Lamarck's notion of transmutation—the belief that giraffes had long necks because their forebears had stretched—Darwin's newest theory accommodated it.

If natural selection highlights all of naturalism's possibilities, then pangenesis illustrates its flaws. One instructive study came from August Weismann, whose mosaic model of development we considered earlier. Weismann had heard stories about a strain of cats with shortened tails, allegedly caused by an unfortunate accident suffered by the group's ancestral mother (folklore held that the wheel of a cart had passed over her tail). This gave Weismann an idea for a mini-evolution experiment of his own. He amputated the tails of a dozen mice, and those of their progeny, for more than five generations, seeking to determine whether this resulted in a new breed of mice with shortened tails. All told, Weismann examined more than 900 mice, yet not one was born with an abnormal appendage. While this did not prove that acquired traits could not be inherited, it deflated whatever enthusiasm remained for Lamarck's theory. Even Darwin came to recognize the flaws in his own model, noting that centuries of male circumcision had not led to the disappearance of foreskin from the Jewish population.

Pangenesis was a worthy attempt, but in the end, it provided an unsatisfactory explanation for hereditary phenomena. Other theories emerged—variations on Darwin's theory that employed new terms (one of which substituted the word "pangene" for gemmule, which may have served as the etymological precursor of *gene*). The idea that traits might possess some physical existence, rather than being embodied in an amorphous vapor, began to gain purchase. But there was no way to define what those physical units were, or how they behaved. The field was at a standstill.

JOHANN MENDEL, COMMONLY KNOWN AS THE "FATHER OF GENETICS," was an unlikely candidate for that title. Growing up on the family farm in Silesia, now part of the Czech Republic, Johann's chores involved tending the garden and the beehives, neither of which he particularly enjoyed. Instead, he was drawn to books, spending large chunks of his childhood in bed reading. The universe, he immodestly believed, held grander plans for him than the world of tilling, planting, and cultivating—a conviction that would prove ironic when his later fame derived from horticultural work.

The Mendels were not wealthy, and Johann's unwillingness to work the fields put a financial strain on the family. If he was going to continue his habit of reading and thinking, his parents declared, then he would have to pay for it himself. At first, the self-confident teen kept himself afloat by tutoring younger students. But eventually, his aspirations outstripped his resources. After considering all his options, he turned to the Catholic Church, hoping that its largess might be able to finance his further (secular) education.

Mendel's gambit paid off. The Abbey of Saint Thomas in Brno—located less than 100 miles from his childhood home—had come to embrace a philosophy emphasizing learning and research over prayer, a practice derived from the Augustinian doctrine of *per scientiam ad sapientiam* (through knowledge to wisdom). Mendel joined the order in 1843, taking his vows with the name Gregor, and found his way into the good graces of Cyril Napp, the monastery's progressive abbot. God's plans for Mendel, Napp believed, did not include leading a congregation, and so he absolved the young friar from the standard rituals—leading prayers, tending to the sick and the poor, and other priestly duties.

For Mendel, who had no desire to become a parish priest, this was an immense relief. Napp continued to look out for the young monk, and after several years he arranged for Mendel to study at the University of Vienna, at the abbey's expense. It was during Mendel's two-year stint there that he first began to think about heredity. His instructors, like all naturalists, emphasized blending. But Mendel was also taking courses in math and physics, more quantitative disciplines that led him to wonder whether heredity was subject to rules—systems like those that governed the properties of elements or the force of a magnetic field. Might there be another way to understand inheritance, he wondered, a mathematically defined association that could predict blended traits with greater precision? Pursuing such a relationship might, at the very least, lead to the breeding of new and useful varieties of plants and animals. But it might also reveal some fundamental principle of nature, like Newton's laws of motion guiding the heavenly bodies.

Upon returning to the abbey, Mendel assumed a part-time teaching position at the local high school (*gymnasium*), but his thoughts kept returning to the question of heredity. As he walked the grounds of the monastery, he became aware of the mice living in its corners and fissures. Some were black, some white, some gray. What conventions, he wondered, govern the choice of hues that decorate a mouse's fur? Might it be possible, through breeding, to discern a quantifiable pattern in coat color? He decided to find out. Ignoring the stench that filled his chamber, Mendel captured and began mating mice of different colors. But before he could make much headway, the bishop—who believed it improper for monks to live in the same space as animals having sexual intercourse—demanded an end to the experiments. Mendel complied, although he noted later that the bishop had ignored one simple fact: plants also have sex.

In retrospect, the bishop's interference was a turn of good fortune. What we know now, but Mendel didn't, is that the coat color of a mouse is a "complex" trait, meaning that it can involve many genes. Had Mendel stuck with mice, the work could very well have led nowhere. Instead, the monk found fresh material for his studies in the five-acre garden on Saint Thomas's grounds, with its hundreds of plant species. Within this botanical playground, the pea plants (*Pisum sativum*) drew Mendel's special attention, for, like the mice, they exhibited distinctive traits that could be easily tabulated. Some were tall while others were short; some produced white flowers while others boasted purple flowers; some made smooth seeds while others made wrinkled seeds. Especially useful was the fact that these traits were "purebred," meaning that a given feature remained true from one generation to the next. Mendel focused on seven purebred characteristics, or *phenotypes*—flower color, seed color, pod color, flower position, plant height, seed shape, and pod shape—and then began to interbreed plants with each trait.

The approach itself was not new. For centuries, farmers had conducted similar crosses, searching for crops with better hardiness, yield, or taste. But Mendel did something that none of the breeders who came before him had thought to do—he counted. How many tall, medium, or short offspring would result from a cross between tall and short plants? How many color schemes could arise from the breeding of white-flowered and purple-flowered parents? And if the offspring of these crosses were bred to each other to yield a second generation—an "intercross"—how would this affect the numbers?

Almost immediately, it was obvious that the blending theory had problems. Instead of yielding a spectrum of outcomes, some traits seemed to override others. For example, crossing a tall plant with a short plant yielded only tall plants, without any of the intermediate heights that blending would have predicted. "Tallness" had some authority, or dominance, over "shortness." Things got even more interesting when Mendel intercrossed the offspring of this first generation, for it was then that the short trait suddenly reappeared in a fraction of that second generation, after having skipped the first. The information prescribing "shortness" had persisted, hidden within the tall parents.

Between 1856 and 1863, Mendel examined some 30,000 plants, scoring their visible phenotypes and looking for patterns. It was an absurd amount of work, but the effort paid off, for a single, striking relationship emerged from all his counting. With uncanny frequency, a three-to-one ratio reproducibly appeared in the second generation of these purebred crosses. For every short plant appearing in the second generation, there were three tall ones; for every white-flowered plant, there were three purple-flowered ones. It made no difference whether the tall plant had white flowers or purple flowers; each of the traits was inherited independently, resulting in the same mesmerizing three-to-one ratio. This was the mathematical precision the monk had hoped to find—a universal logic to inheritance. The challenge now was to make sense of it.

FORM-BUILDING ELEMENTS

OVER TIME, MENDEL FORMULATED AN ALGEBRAIC MODEL WITH which to explain the patterns. The data, he deduced, were most consistent with traits being defined by binary units—what he called *bildungsfähigen Elemente*, or "form-building elements." These elements, he speculated, were supplied in equal quantity by each parent—one from the male (pollen) and one from the female (ovule). Mendel reasoned that the elements could assume either of two forms (or *alleles*): a "dominating" one, like tall height or purple color, or a *recessive* one, like short height or white color. For ease of notation, he used capital letters to denote *dominant* alleles (*T* for tallness) and lowercase letters to denote recessive ones (*t* for shortness). A cross of a purebred tall plant (*TT*) to a purebred short plant (*tt*) would result only in *Tt* offspring, and because *T* was dominant to *t*, this entire first generation would appear tall. But if *Tt* plants were crossed to other *Tt* plants (i.e., the second generation), things got more complicated, as the alleles could combine randomly in four different ways: *TT*, *Tt*, *tT*, and *tt*. Of these, three combinations (*TT*, *Tt*, and *tT*) would result in tall plants because of the dominance of the *T* allele, while only one combination with two recessive *t* alleles (*tt* plants) would result in small plants. Hence the recurring three-to-one ratio.

Mendel's form-building elements—what we now call genes—were the missing link in evolutionary theory, the insight that made the model of natural selection complete. But the message didn't reach its intended audience (Darwin) during Mendel's lifetime. In 1865, he presented his theory in two lectures in Brno, laying out the mathematical evidence for dominant and recessive alleles governing traits. But it is likely that those in attendance didn't understand that they were witnesses to a new theory of heredity; instead, they had simply heard a monk speaking to them about gardening. The following year, Mendel published his work in an obscure journal, the *Proceedings of the Natural History Society of Brünn*, with the less-than-catchy title "Experiments in Plant Hybridization." It vanished from view almost immediately. Fewer than a handful of schol-

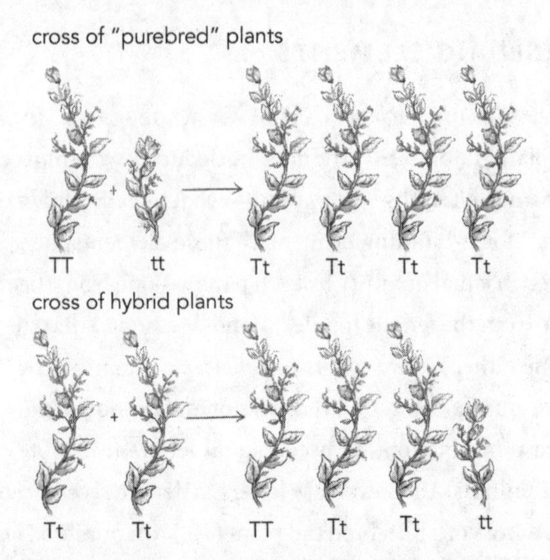

A Mendelian cross between purebred tall pea plants (genotype *TT*) and short pea plants (genotype *tt*). The *T* allele is dominant to the *t* allele. Consequently, the hybrids resulting from this cross—carrying one tall allele and one short allele (genotype *Tt*)—are all tall. An intercross of the hybrid offspring produces both tall and short plants in the second generation, with each plant's height depending on its genotype.

ars ever saw the manuscript, and most of those who read it considered the work an exercise in plant breeding, not the comprehensive account of inheritance it represented. From the dozen or so scientists to whom he sent reprints, Mendel received only a single, disparaging, reply. For all their importance, the monk's findings remained unseen for decades, a misplaced masterpiece collecting dust on library shelves.

THE RECOGNITION MENDEL HAD IMAGINED FOR HIMSELF CAME ONLY after his death, when three botanists, working some 30 years later, found the same striking three-to-one ratios in their plants and subsequently stumbled upon his little-known manuscript. Much had changed in the decades since Mendel's initial report, and what had been an unreceptive scientific world in 1865 was now open to the possibilities raised by the prospect of invisible yet mathematically defined units of heredity. New ideas blossomed, the most exciting of which related to the still-unfilled gap in Darwin's theory. Biologists had continued to pick at this missing link—wondering how variants arise in the first place and how they get transmitted to offspring—and Mendel's model provided a plausible answer.

One of the champions of this view was Hugo de Vries, the most respected botanist in Holland at the turn of the twentieth century and one of the "rediscoverers" of Mendel's paper. De Vries didn't work with peas, but instead studied the evening primrose (*Oenothera lamarckiana*). In contrast to Mendel, who scrutinized variation that was already present in the garden when he began his work, de Vries claimed, amazingly, to have observed new traits arising out of nowhere. He referred to these emergent features, a change in the shape of a leaf or the size of a plant, as "monstrosities," a term he later changed to "mutations." Importantly, such variants were not "one-offs": de Vries observed that the new traits could be passed along to subsequent generations, making them, in effect, new organisms. Mendel's form-building elements, the original name for genes, might govern more than the simple inheritance of traits—they might also govern the emergence of new ones.

THE DISCOVERY OF *WHITE*

WHEN WE LAST ENCOUNTERED THOMAS HUNT MORGAN, THE YOUNG embryologist from Kentucky, he had decided to move to Europe, where the most exciting biology seemed to be happening. In 1894, he took a leave of absence from Bryn Mawr College and set sail for the Stazione Zoologica in Naples, where scientists like Hans Driesch, the embryo shaker, were pushing the scientific envelope. Driesch was still hard at work probing plasticity, the embryo's ability to change course, and he and Morgan became fast friends. With Driesch as his tour guide, Morgan quickly acclimated to his new environment, studying cell division by day, and exploring the sights, smells, and sounds of Naples by night. Slowly, Morgan's scientific training as a naturalist gave way to a deeper level of analysis—a transition from observation to perturbation, from organism to cell, from taxonomy to mechanism.

In 1904, Morgan joined the zoology department at Columbia University. His teaching responsibilities included talking to graduate students, and one day a student named Fernandus Payne showed up in his office with an idea for a project: What would happen to an animal bred for multiple generations in the dark, Payne wondered—would it become blind? The idea held little appeal for Morgan, who thought it held little chance of success (plus, it bore too close a resemblance to Weismann's attempt to force heritable changes in mouse tails). Instead, what caught his attention was the organism Payne proposed using for the study—the fruit fly, *Drosophila melanogaster.*

The thought of working with fruit flies hadn't crossed Morgan's mind during his time in Europe, but he had been hearing more and more about these insects lately. Easy to maintain and subsisting on a diet of decomposing fruit or vegetable, the insects reached sexual maturity within two to three weeks of birth and bred by the hundreds. While Payne sought to exploit these properties to fit his breeding strategy (10 generations of darkness) into a single semester, Morgan had other ideas. Specifically, he was thinking about the newly observed strains of evening primrose that de Vries had claimed to observe—the so-called monstrosities—and wondered whether a similar phenomenon might be observable in animals. *Drosophila*, with its ability to reproduce quickly, would make such a mission feasible, as tens of thousands of creatures could be examined in matter of weeks. With the encouragement of his colleagues, and flies from Payne, Morgan embarked on a multiyear search for a one-in-a-million fly. He could not predict what such a fly would look like, only that it should be different, in some way, from all the others.

The Fly Room was born.

"WELL, HOW IS THE WHITE-EYED FLY?" MORGAN'S WIFE ASKED FROM her bed.

It was January of 1910, three days after the birth of the couple's third child, and his wife was still in the hospital recuperating. Morgan proceeded to recount the latest news from the lab when suddenly, remembering where he was, the biologist paused. "And how is the baby?" he finally asked.

Morgan had been preoccupied with the discovery, just a few days earlier, of the special fly he had been waiting for. In one of the milk bottles used to hold the colonies of fruit flies being investigated, he had seen a singular animal navigating its way among the scraps of rotting food and the dozens of other flies trapped inside. What made it special was that in place of the brilliant red orbs that characterized the eyes of a normal fly, this one had vacant pale spheres, devoid of color. Despite seeing the specimen with his own (human) eyes, it was some time before he was prepared to trust his senses.

The quest for a mutant fly had been more arduous than Morgan expected. For years, he had bred the insects with nothing to show for it. So when an outlier suddenly appeared, distinct from the thousands of other flies he had scrutinized, Morgan could scarcely believe it. He examined the insect daily looking for hints of red, which might indicate a developmental delay rather than an absence of pigment—a trivial explanation. The phenomenon he was searching for was not merely a delay; he sought a change that was durable and heritable. As time passed and the fly's eyes remained pallid, Morgan slowly began to accept his good fortune. With little fanfare, he gave his newfound specimen, and the gene he believed responsible, the nickname "*white*."

Determining whether the white-eye phenotype was heritable required mating. Carefully, so as to not lose the precious specimen, Morgan allowed the *white* mutant fly to breed with normal, red-eyed flies (so-called "wild-type" animals). To his dismay, all the flies in this first-generation cross, the so-called F1 generation, had red eyes. But recalling how the recessive traits of Mendel's pea plants—short height and white flowers—had appeared to "skip" a generation, he carried on. His disappointment turned to delight with the second, or F2, generation, for when Morgan allowed the F1 flies to interbreed, the offspring gave exactly the results that Mendel would have predicted: for every three flies with red eyes, there was one whose eyes were as pale as the moon.

Morgan drew two conclusions from these results. First, heritable new traits—mutations—could arise spontaneously in animals, just as de Vries had shown for plants. Second, it appeared that plants and animals followed the same rules of inheritance, with dominant and recessive alleles. Still, there was something odd about the pattern of inheritance—an unexpected relationship between a fly's eye color and its sex. Specifically, *all* the females from the F2 cross had red eyes, while *half* of the male flies from the cross had white eyes. (It was only when males and females were combined that the three-to-one ratio was apparent.) The trait behaved like one of Mendel's recessive alleles, but in an unusual manner that seemed unique to animals. Somehow, sex was limiting the inheritance of *white*. But how?

MAPPING GENES

IN THE ERA OF BLENDING, HEREDITY HAD AN ETHEREAL QUALITY, A disembodied character that could reside either inside or outside of the cell. But these newly recognized hereditary elements, genes, were different—quantifiable, measurable, and predictable. Once it was understood that genes constitute the units of inheritance, the previous flimsiness of thought was no longer suitable. Scientists could begin to think about heredity, evolution, and development in concrete terms, the output of some yet-to-be-discovered chemicals in the cell. Genes, one could say, must have a home.

Small cellular fragments called chromosomes, located in the cell's nucleus and invisible under normal circumstances, constitute that home. Named for their radiant properties when stained with colored "chromophore" dyes, chromosomes become visible during cell division, condensing into tiny specks resembling chipped pencil graphite. Microscopists had observed these particles since the early days of cell theory, but their function, like that of most cellular components, had been unknown. Walter Sutton, an American graduate student studying these mysterious particles at the turn of the twentieth century, observed that almost all cells contained an even number, paired together like animals on Noah's ark. But there was an exception. Gametes—an animal's eggs and sperm—contained only a single, unmatched, set of chromosomes. Upon fertilization, when egg and sperm fused, the zygote once again recovered the pair. Struck by how similar this pattern of separation and rejoining was to Mendel's now-rediscovered principles of inheritance, Sutton hypothesized that chromosomes were the physical conveyors of Mendel's dominant and recessive alleles.

At the same time, German scientist Theodor Boveri, another veteran of the Stazione Zoologica, was reaching similar conclusions. Using Driesch's favorite animal, the sea urchin, Boveri managed to obtain eggs that had been fertilized with more than one sperm. The resulting zygotes had too many chromosomes, and they invariably died prematurely. But every so often, a newly fertilized egg discarded the extra chromosomes, and when this happened, the sea urchins looked normal. Working independently, Sutton and Boveri proposed the comprehensive theory of inheritance that we have since adopted as canon. Chromosomes, those tiny specks of matter, are the physical incarnations of heredity—matched pairs constituting the blueprints for creating an organism. Moreover, their research showed that embryos have a means of counting chromosomes, and only those with the correct number develop normally.

MORGAN WAS AT FIRST DISMISSIVE OF SUTTON AND BOVERI'S CHROMOSOMAL theory of heredity, flippantly referring to its adherents as those "chromosomal people." But the discovery of *white* got him thinking: Could Boveri and Sutton have been right? Might chromosomes have something to do with the strange association between a fly's eye color and its sex?

A possible answer came from Nettie Stevens, one of Morgan's former students. Like her mentor, Stevens had made a pilgrimage to the Stazione Zoologica in Naples, where she overlapped with Boveri. Inspired by his chromosome work, Stevens returned to Bryn Mawr, where she established a research program focused on these insect units of inheritance. Building on prior studies showing that male and female insects have a different chromosomal makeup, Stevens focused on the "X element," a chromosome that was typically present in two copies in females and one copy in males. Observing that many male insects carry a smaller "odd chromosome" to accompany their single X element, Stevens correctly theorized that this anomalous chromosome—later called Y—was responsible for determining male sex.

Morgan saw that Stevens's theory might explain *white*'s odd pattern of inheritance. Investigating his *Drosophila* in greater detail, Morgan found that the insects' chromosomal makeup correlated with sex just as Stevens had suggested: female flies had two matched X elements, or chromosomes, while males had a mismatched X and Y. This, in turn, led to Morgan's greatest insight: What if the hereditary determinant responsible for eye color was physically connected to the X chromosome? If so, it would explain why males, with their single X, were prone to develop white eyes, while females, with their two Xs, were not.

Let's break this down a bit more. Imagine a male fly carrying the recessive *white* allele (*w*) on its single X chromosome, and let's denote this genetic constitution, or *genotype*, as X^wY. Such a fly would be expected to have white eyes because of the unopposed activity of the mutant form of the gene. By contrast, if a female fly inherited the *white* allele, it would still carry a normal allele (+) on its other X chromosome (here, we would denote the genotype as X^wX^+). Because of the recessive nature of the *white* mutation, the normal allele would supersede the activity of the mutant allele, conferring red eyes to the X^wX^+ female.

To convince himself, Morgan continued his breeding long enough for female flies to inherit two mutant Xs by chance (X^wX^w). Now, white-eyed females began to appear. The interpretation was astonishingly simple and yet still revolutionary: the information encoding the white-eyed trait was located *within* the substance of the X element. Whether one wished to call it a determinant, a form-building element, or a gene, it was a physical embodiment of a heritable directive. From that moment on, the trait-determining units of heredity were no longer abstract factors or amorphous particles. They had a physical presence, a chemical existence, residing somewhere in that pigmented matter known as a chromosome.

IN THE MONTHS THAT FOLLOWED, MORGAN'S TEAM IDENTIFIED MORE than a dozen new mutations in flies—mutations that caused misshapen wings or unusual body color. Although *white* had taken years to find, it was surprisingly easy to identify new mutants if one knew how to look for them. Genetics, as the nascent field was being called, swept across the scientific world.

As the Fly Room scientists probed these mutants in greater detail, interesting and unexpected inheritance patterns began to emerge, patterns that deviated even further from Mendel's simple predictions. For example, consider two other recessive mutant strains that Morgan discovered. One strain, *vermillion* (*v*), bore brilliantly colored scarlet eyes. Another strain, *miniature* (*m*), had stunted, misshapen wings. Bred on their own, each strain behaved similarly to *white*, giving rise to mutant phenotypes at the expected ratios (and with the same pattern of sex-associated inheritance). The unexpected result came when Morgan tried to generate flies carrying both traits at the same time, an effort that proved extremely difficult.

This result was surprising and quite different from anything Mendel had observed in peas, whose traits all appeared to be independently inherited. Mendel could manufacture plants having any combination of height, flower color, or seed shape he desired. But something else was going on here—a force that prevented the genes controlling these traits from behaving independently. To Morgan, with his understanding of the physical nature of heritable information, this could mean only one thing: the genes encoding these traits were physically connected, or linked, on the same chromosome.

Alfred Sturtevant, a Columbia undergraduate in his junior year, took a more careful look at the data. Sturtevant noticed that while the offspring of Morgan's crosses had either vermillion eyes *or* miniature wings most of the time, a fly having both traits did emerge on rare occasions. Sturtevant found other traits whose chances of being inherited together also deviated from the Mendelian expectation. The frequency of these deviations varied from trait to trait—occurring anywhere from 1 to 10 percent of the time. Sturtevant likened these linked genes to passengers on a train, seated in either the same car or cars that were widely spaced apart. As the train proceeded on its hypothetical journey, its carriages might be periodically separated and swapped. If this happened, passengers seated several cars apart would have a good chance of finding themselves on different trains. By contrast, passengers seated in the same car would be sure to end up at the same destination. In other words, the likelihood of two travelers parting ways or sticking together would depend on the distance between their seats.

Applying this logic to inheritance, Sturtevant realized that the closer two genes were to each other on a chromosome, the less likely it was for them to be inherited separately. Indeed, the only way for this to happen, giving rise to a new combination, would be a physical reconfiguration of the chromosome. Genes located far apart, or on different chromosomes, wouldn't face this problem—they would be inherited in more-or-less random combinations according to Mendel's doctrine. But genes located close to each other on the chromosome would be unable to be distributed so freely. The difficulty in generating flies with vermillion eyes and miniature wings was because the genes encoding these traits were too close together!

In a single night, to the neglect of his homework, Sturtevant charted the relative positions of six genes on the X chromosome based on the likelihood of any two linked genes becoming separated, creating the first "genetic map." In the years that followed, more detailed maps emerged from the Fly Room, recording the relative positions of dozens of *Drosophila* genes. But while the work answered one question—Where are genes located?—it raised a new one: What are they made of? What is the chemical substance within chromosomes that dictates color and shape and distinguishes a sea urchin from a human being? It was a question that could be answered only by chemists, with a scientific lineage completely detached from the genetic tradition of Mendel, de Vries, and Morgan.

A SUBSTANCE LACKING SULFUR

IN THE LATE 1800S, IN A CASTLE OVERLOOKING TÜBINGEN, GERMANY, a young physician named Friedrich Miescher was studying pus. The 25-year-old Miescher came from a line of Swiss physicians, and his parents had groomed him to enter the family profession. Miescher went through all the motions, up to and including passing his board examinations, but the more time he spent in practice, the more he grew weary of the patients—listening to their complaints and offering advice but rarely effecting a cure. More exciting to him was a new frontier in chemistry: the application of chemical principles to the actions of cells, which many believed was the best way to advance knowledge in human physiology. In 1868, Miescher packed up his things and moved to Tübingen, where he became apprentice to the famed chemist Felix Hoppe-Seyler.

By the 1860s, cell theory—the notion that the cell is the basic unit of every tissue and organism—had achieved the status of dogma, and this led researchers to look further and deeper inside the cell. This revealed a whole world within—subdivisions, or organelles, whose functions awaited clarification. Many researchers, applying the standard naturalist approach, spent their time characterizing these minuscule structures and speculating about their purpose. But a small number of chemists—among them Hoppe-Seyler—set out to determine the precise molecular composition of these subcellular zones.

Hoppe-Seyler's laboratory, with its vaulted chambers, was situated in what had once served as the castle's laundry, its tubs replaced with broad tables and its washboards exchanged for flasks, beakers, stirrers, and heating elements. In cabinets lining the walls were the reactive chemicals that could tear cells and tissues apart, reducing them to their constituent elements. Added in the proper concentrations, combinations, and sequences, and heated to a desired temperature, these materials—acids, alkalis, solvents, and alcohols—could expose the molecular makeup of any substance. It was a means of dissecting human tissue, using molecules and reactions instead of scalpels and scissors.

At his mentor's suggestion, Miescher obtained discarded surgical dressings from a nearby clinic and isolated cells from the pus-laden bandages. This suited his purpose well, as pus is saturated with white blood cells, or leukocytes (*leuko*, white; *cyte*, cell), and Miescher could easily separate them from the purulent starting material. Hoppe-Seyler had a seemingly simple assignment for his young apprentice: define the chemical nature of the nucleus, the biggest organelle in these disease-associated cells.

Miescher proceeded, as chemists do, with a systematic series of extractions, purifications, and reactions. First, he needed to separate intact cells from debris, a relatively easy task that involved discarding anything that showed signs of decomposition. The next step required separating the nucleus from the rest of the cell. This had never been done before, but Miescher found that a weak solution of cold hydrochloric acid broke the cells open just enough for most of their contents to float away while the nucleus remained intact. Finally, Miescher treated his samples with an agent isolated from the stomachs of pigs called pepsin, which seemed to remove any remaining contaminants. Inspecting the purified nuclei under the microscope, Miescher felt confident that he had a clean sample to work with, and he proceeded with his chemical dissections, ready to lay bare the material within.

FULLY UNDERSTANDING A SUBSTANCE AT THE CHEMICAL LEVEL entails reaching three experimental milestones: defining its composition, solving its structure, and achieving its synthesis. The first of these, determining composition, is a brute-force effort—a trial-and-error exercise of chemical demolition that reduces a substance to its component elements. The second milestone—determining structure—is a different matter. Compounds made of the same elements may take on vastly different forms depending on how those elements are pieced together. Chemical structures are the chemist's acrostic puzzles; some can be deduced intuitively while others require a squadron of supercomputers. Finally, synthesizing a compound—making it from scratch from its component parts—is the chemist's ultimate measure of success.

Miescher began to apply all the chemist's tricks to his newly purified nuclei. Some chemicals caused the nucleus to rupture, resulting in a viscous solution. Others caused a cotton-like substance to materialize, fine strands of snow-white thread that became more and more defined with time. Miescher proceeded through his chemical checklist, subjecting the nuclear-derived material to increasingly harsh conditions to determine the elemental composition of this new substance, which he had come to refer to as "nuclein." The substance contained elements that were shared by the other cellular components that Hoppe-Seyler and fellow chemists had found in living tissue. These elements—carbon, oxygen, and nitrogen—seemed common to most biological materials, including protein. But nuclein stood apart, in that it contained an overabundance of phosphorus but virtually no sulfur.

Miescher published his findings in 1870 with his adviser's blessings. He had a sense, biased as it was, that there was something special about the material he had purified—that one day it would turn out to be every bit as important as protein. But the enthusiastic reception he hoped for never came, and Miescher would suffer the same fate of obscurity Mendel had. It was only decades later that nuclein—the first crude preparation of DNA—would get its due.

BY 1920, IT WAS AGREED THAT GENES RESIDE IN THE NUCLEUS OF THE cell, assembled on chromosomes in linear fashion as Morgan and Sturtevant had concluded. Why, then, was nucleic acid (as nuclein had been renamed) not the obvious candidate for the genetic material? The answer, ironically, was that it was deemed too bland.

Following Miescher's initial characterization, analysis of nucleic acid revealed that the substance comprised three molecular components: phosphoric acid, the source of the substance's high phosphorus content; a sugar, deoxyribose; and one of four "bases"—adenine (A), guanine (G), thymine (T), or cytosine (C). Scientists had begun to speculate about the chemical structure of nucleic acid—how these building blocks might be arranged in three-dimensional space. One of them, a Russian-born physician named Phoebus Levene, concluded that nucleic acids were merely repeating elements of the four bases, a random mixture of A, G, T, and C assembled against a sugar-and-phosphate "backbone." Viewed in that way—with a presumed monotonous structure—nucleic acid seemed more likely to serve some mechanical function within the nucleus than to be capable of bearing hereditary information. Levene's prominence gave the model credibility, and it took nucleic acid out of the running as the genetic material.

Proteins, by contrast, seemed to be much more promising candidates. Composed of a 20-letter alphabet of "amino acids"—a far richer palette than the four bases of nucleic acid—proteins are the most abundant macromolecules in a cell. A protein's sequence, and hence its structure, is defined by the unique order of its amino acids, resulting in a vast amount of combinatorial information. For example, a protein having only five amino acids represents three million possible combinations. As most proteins in nature contain hundreds of amino acids in sequence, the number of "words" that can potentially be encoded is immense, a diversity that allows proteins to assemble into a vast array of shapes, sizes, and functions. Given this immense potential for variation within the protein universe, there was no pressing need to consider other candidates—especially nucleic acid, with its seemingly innocuous structure—for the hereditary substance. Nucleic acid would have its heyday soon enough. But for the time being, the biochemists seemed content to follow the Mark Twain dictum: Never let the truth get in the way of a good story.

THE TRANSFORMING PRINCIPLE

THE SPANISH FLU PANDEMIC OF 1918 KILLED AN ESTIMATED 50 TO 100 million people, 3 to 5 percent of the world's population at that time. Most of the deaths were due not to the flu virus itself but rather to bacterial infections that took advantage of the lungs' weakened defenses. In the years following the devastating outbreak, British bacteriologist Frederick Griffith began to study one of the bacteria responsible, a common inhabitant of the respiratory tract called *Streptococcus pneumoniae*, or "pneumococcus." Griffith took samples from pneumonia patients, and after culturing the bacteria in the laboratory, he introduced them into mice.

The cultures fell into two groups: an "R" form, which caused minimal symptoms, and a virulent "S" form, which typically killed the host. In a clever series of experiments, Griffith found that he could transfer the lethal properties of the virulent strain to the benign strain by mixing them together. This transformation occurred even if he heat-killed the virulent bacteria before mixing, indicating that a resilient substance in the lethal strain—what he called the "transforming principle"—was responsible for the genetic transmission of the trait.

To Oswald Avery, a physician-scientist at the Rockefeller University in New York City, Griffith's observations had implications that extended far beyond the world of bacteriology. Avery, who had also studied pneumococcus, was a fan of Griffith's work, and in the early 1940s, on the verge of retirement, he decided to exploit the British scientist's findings to understand the chemical basis of heredity. Was it a protein that could transform the cells from benign to lethal? If so, what was the nature of this special protein? Or was it some other substance, an unforeseen trigger, that set the bacteria's deadly behavior in motion?

Enlisting the help of two colleagues—Colin MacLeod and Maclyn McCarty—Avery began his quest to identify the causative agent. He split open cells from the virulent S strain and separated them into their chemical components—proteins, carbohydrates, lipids, and nucleic acids. Then, after incubating each "fraction" individually with a culture of bacteria from the R strain, the trio noted which, if any, could transform the benign bacterium into a more virulent form. Contrary to the expectations of the scientific world, and those of the three Rockefeller scientists, the protein fraction was unable to confer the lethal trait; instead, that property resided with the nucleic acid fraction. Previously ruled out as a mediator of heredity on the grounds that it was too simple, the substance known by its chemical shorthand—deoxyribonucleic acid, or DNA—spelled the difference between life and death.

IN THEIR 1944 PAPER DESCRIBING THESE RESULTS, THE ROCKEFELLER scientists declared that Griffith's transforming principle consisted "principally, if not solely," of DNA. But their findings clashed with the scientific canon, and they met major pushback. Scores of scientists had staked their careers on the belief that genes are composed of proteins, and they were not going to accede quietly. Most complaints revolved around the possibility that Avery's nucleic acid preparation might have still contained some protein—a tiny bit of contaminating material that could be responsible for the lethal transformation. Avery, MacLeod, and McCarty did everything they could to convince the naysayers, including a demonstration that enzymes which chew up proteins (proteases) had no effect on the transmission of virulence. But the critics held fast, wedded to the idea that genes were made of proteins.

Ultimately, it was virology—not biochemistry—that quelled the skeptics. At Long Island's Cold Spring Harbor Laboratory, less than 50 miles from Avery and his team, Alfred Hershey and Martha Chase were studying *bacteriophage*—viruses that infect bacterial cells. Like all viruses, *phage* consist of a nucleic acid core packaged in a protein shell. When phage infect a host, they latch on to the surface of the bacterial cell and "inject" their genetic payload inside. But because viruses are incapable of reproducing on their own, phage must hijack the already-existing machinery for reproduction within the host, which in turn sets off a cascade of viral replication.

Hershey and Chase devised a clever method to determine which component of the virus—its DNA or its protein—was responsible for viral propagation. In one experimental setup, they allowed phage to replicate in the presence of radioactive sulfur, causing the virus's proteins to be "labeled." (DNA, as a substance lacking sulfur, remained unlabeled under these conditions.) Then, in a separate experiment, they grew the virus in the presence of a radioactive phosphorus, causing the DNA to be radioactively labeled. (Under these conditions, some proteins would also be labeled, but the majority of the radioactivity would be derived from the DNA.) Hershey and Chase then mixed each of these viral preparations, separately, with fresh bacteria. After enough time had passed for the viruses to infect the cells, the researchers simply looked to see whether the bacterial cells contained labeled sulfur, labeled phosphorus, or both. The answer could not have been clearer: phosphorus was the only radioactive substance to enter the cells; the radioactive sulfur remained on the surface, but never got inside.

DNA, and DNA alone, held all the instructions for making a new virus.

BEFORE UNDERTAKING HIS GENETIC STUDIES, MORGAN HAD MISGIVings about Mendel's theory, believing that animal development was too complex to be determined by the kinds of form-building elements that governed the height of a simple pea plant. His instincts were apt, as most animal traits are not determined by a single gene. It is thought that more than a hundred genes influence human height and more than a thousand influence intelligence, each contributing a small part to the overall outcome. But mutations affecting single genes can also have dramatic effects. A misplaced C or an erroneous T—typos affecting a solitary letter out of the billions that make up the DNA of a human cell—can, if it occurs in the wrong place, change the meaning of a word, resulting in devastating illnesses or congenital abnormalities.

Genes can act by suggestion or by decree.

The sum of an organism's DNA is referred to as its *genome*. Genome size varies from creature to creature, but it bears surprisingly little relationship to organism complexity. The human genome contains just over 6 billion bases, divided equally between the two copies of our 23 chromosomes. By contrast, some fish have genomes that are 10 times that size (with little to suggest that those fish are more complex than human beings). In most genomes, including mammals', only a fraction of the DNA consists of protein-coding genes, while the remainder—what used to be called "junk DNA"—occupies the space in between. Consequently, the size of a genome does not predict how many genes it contains or how the organism uses them. For example, the sea urchin genome, with its 800 million bases, is about a tenth the size of the human genome but contains roughly the same number of genes: 20,000 to 25,000.

The recognition that DNA is the genetic material revolutionized biology, but it took longer than it should have. The fact that nucleic acid was dismissed as too simple to play that role stemmed from the failure to consider that the *order* of DNA's bases, rather than its overall composition, was critical—the type of impediment to understanding that the biologist Stephen Jay Gould calls a "conceptual lock." This lock was finally broken in 1953, when James Watson and Francis Crick revealed their now-famous double-helix model, showing how DNA's four-letter alphabet of G's, A's, C's, and T's could be arranged in different ways to spell an almost infinite number of genetic words.

What started with studies of peas and pus, and gained momentum

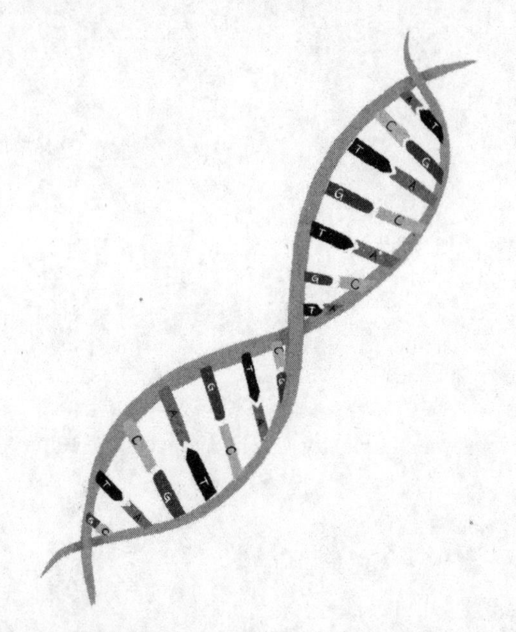

The two strands of DNA wind around each other like a twisted ladder, its rungs formed by bases with a chemical affinity for their partners: adenine (A) for thymine (T), and cytosine (C) for guanine (G).

with flies and phage, had by the mid-1950s blossomed into a foundational perception of inheritance at the cellular and organismal levels. Genes—those parcels of hereditary information first parsed out by Mendel—were responsible not only for the emergence of new species during evolution but also for guiding the formation of a new organism during development. The challenge awaiting the next generation of scientists was to understand the inner workings of this genetic machinery, and how its dictums managed to produce an adult animal from a single cell.

Chapter 3

SOCIETIES OF CELLS

Presume not that I am the thing I was.

— WILLIAM SHAKESPEARE, *HENRY IV, PART 2*

Imagine that you have made one of the great breakthroughs of the twentieth century, but you can't get anyone to believe you. When you were a student, nobody wanted you to follow your passion (which happened to be insects). Nobody wanted you to go into biology. Nobody really cared what you were working on. And now that you have accomplished something important—obtained a result that no one previously expected—it is almost impossible to get anyone to have confidence in your result. As far as your senior colleagues are concerned, you have no credentials and no experience. You can't be taken seriously. You are nobody.

In the fall of 1957, in a small laboratory in Oxford University's Department of Zoology, a soft-spoken graduate student named John Gurdon faced just such a predicament. This was a time of exponential discovery in biology. Only a few years had passed since Hershey and Chase proved that DNA is the bearer of genetic information and Watson and Crick put forth that the molecule exists in a double-stranded double-helical structure. Those discoveries had launched a race to understand the syntax and grammar of this newly uncovered language of heredity. But for Gurdon, who had gotten into graduate school by the skin of his teeth, these matters were secondary. To be sure, he was curious about all the molecular advances happening around him—discoveries that revealed how cells read, copy, and repair their DNA. But these discoveries struck him as fine brushwork on a large canvas, and he was more interested in the spaces on the canvas that were still blank.

Gurdon was a developmental biologist, fixated on the One Cell Problem—the amazing fact that animals as diverse and complex as earthworms, kangaroos, and human beings all arise from a single microscopic cell. It was a question the molecular biologists, with their frenzied detail work, did not fret over much. Gurdon was interested in the bigger picture, in understanding the principles by which genes collaborate to form an organism. This quest would ultimately lead him to be the first person to *clone* an animal from a mature cell, work that established the basic relationship between genes and development. He would eventually win biology's highest awards. But before any of this could happen, he needed to convince a skeptical world that he hadn't made a mistake.

DECISIONS, DECISIONS, DECISIONS

EVERY ORGAN IN THE BODY IS A SOCIETY OF CELLS, ITS CITIZENS EACH with their own defined positions and roles to play in a larger social order. The corporeal empires that make up our organs are built slowly during development, and every trade, as it were, is represented within them. There are cells whose jobs are mundane and repetitive—red blood cells that carry oxygen around the body; muscle cells, or myocytes, that bear colossal loads; and keratinocytes who sacrifice their lives to create the protective shell of the skin. It is these "blue collar" laborers who give each tissue its function and character. Skilled workers are represented as well: osteoblasts and osteocytes in our bones that act as construction (and demolition) crews, enterocytes that extract nutrients from the diet, and endocrine cells whose hormones synchronize the above activities across tissues. These cellular laborers, in turn, are overseen by a managerial class—neurons who measure output and establish rates of pro-

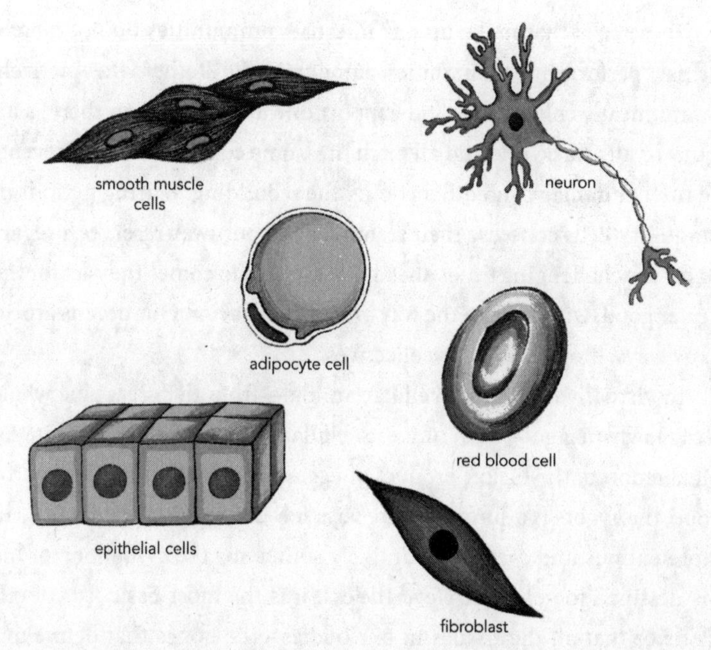

The many types of cells that make up our tissues have different structures that subserve their functions. Shown here are a group of smooth muscle cells (which cause tissues like the gut tube and the diaphragm to contract), an adipocyte (which stores fat), a neuron (which transmits signals within the nervous system), a red blood cell (which carries oxygen throughout the body), a layer of epithelial cells (which serves as a barrier to the outside world), and a fibroblast (which forms the body's connective tissues).

duction as needed. Like all societies, the cellular collectives that make up our tissues harbor delinquents—rogue cells that can grow out of control into life-threatening cancers, or vigilantes whose misguided escapades wind up causing scars. To deal with these, our tissues also employ first responders—neutrophils and macrophages, which extinguish fires before they get out of control, and lymphocytes, which constitute an aggressive border patrol prepared to evict (or execute) unwelcome trespassers on sight. Some cellular citizens, like stellate cells, hold down two jobs, taking leave of their normal trade of storing vitamin A when called upon, assuming the auxiliary task of laying mortar to help fortify an injured tissue.

These cells that make up our internal communities do not merely coexist, performing their duties autonomously. Rather, they actively communicate, collaborate, and support one another. When there is an injury or an attack, one and all pitch in—some concerned with preventing further damage and others doing the rebuilding. During periods of tranquility, cells conserve their resources and put away reserves, preparing for the challenging times ahead that are sure to come. They act for the greater good, with little in the way of selfish behavior. Our organs are, in many ways, utopian cellular collectives.

In virtually every multicellular animal—from the great blue whale to the lowly maggot—each of these cellular societies begins as a single-celled embryo, the fusion product of egg and sperm. Under the microscope, the zygote is a simple sphere with few distinguishing features, an unpretentious and seemingly unlikely source for the symphony of life it is destined to generate. Nevertheless, it is the most basic fact of our existence that all the tissues in our bodies—the bones that define our shape, the muscles with which we manipulate our limbs, and the brains that allow us to ponder our very existence—all trace their ancestry back to this single primordial cell.

The challenges the embryo faces as it makes the trip from zygote to adulthood are formidable. A newborn human infant contains more than a trillion cells. An adult has more than 10 times this number. To achieve this size, the zygote divides with metronome-like precision, reiterating the process of replication more than 40 times prior to birth. As their numbers grow, cells begin to diversify, undergoing a process of *differentiation* during which they shift from free agents to dedicated laborers. Differentiation is what gives our cellular societies their distinctive occupational makeups, ensuring that each organ has the right combination of workers, suppliers, managers, and defenders. But for the individual cell, differentiation represents a trade-off, for when a cell commits to one specialized path, it closes the door on other options. A cell that has committed itself to become part of the heart forgoes its opportunity to join the kidney, and vice versa.

Differentiation is something of a black box of development. By the time the embryo has accrued more than a dozen cells, only those that linger on the surface can still be seen; the rest, buried inside, are out of reach. Thus, the events that guide a cell's choice of occupation are concealed, teasing anyone wishing to unscramble the mysteries of cell specialization. And there is also a more profound obstacle: even if it were possible to peer into the embryo's interior, this would provide little information about choices of cellular career. Developing cells do not announce their chosen professions ahead of time, declaring "I will be cartilage" or "I will be skin." Differentiation makes itself apparent only in hindsight, well after the decision to specialize has already been made.

In vertebrate animals, once the zygote has divided several times—into two cells, then four, then eight, and so on—it forms a hollow ball referred to as a blastula (or *blastocyst* in mammals). At this incipient stage of life—transpiring within hours to days of amphibian or mammalian development—embryonic cells are somewhat interchangeable, not yet committed to one occupation or another. They are the embryonic equivalents of infants, having all potential fates before them, and it was this plasticity that Driesch and later Spemann happened upon, as we saw in Chapter 1. But the period that follows—the tumultuous adolescence of the embryo—is the stage, called gastrulation, that marks the period in development when differentiation begins in earnest. It is also a time when developmental mechanisms become particularly hard to tease apart.

The process of differentiation gives diversity to our cellular societies. The question, then, is: How do our motley cellular collectives come into being? What instructions tell some cells to become the absorptive enterocytes of the intestine, others to become the oxygen-carrying erythrocytes of the blood, and yet others to lay down the bony matrix of the skeleton? Under what mysterious influences do cells make these kinds of decisions? Or are the decisions made for them, molecular assignments by some senior authority, a cellular hiring officer that fills open positions as needed and directs new personnel to their correct positions in the body?

GURDON'S PATH TO SCIENCE WAS CIRCUITOUS. DURING HIS BOYHOOD in a small English village in Surrey, insects had fascinated him. He fell in love with the endless variety of patterns on moths and butterflies that abounded in the countryside near his home. Hours melted away as he pored over his specimens, growing caterpillars to watch their metamorphosis and categorizing them by species with a dense textbook of entomology that weighed nearly as much as he did. There seemed to be no limit to the diversity of these creatures, nor any end to the questions that could be asked about them. Already an expert lepidopterist, he could imagine nothing more satisfying than a life studying these remarkable beings.

But his timing was off. Born in 1933, Gurdon came of age in the aftermath of World War II, a time when England's secondary school system had little patience for curiosity or challenges to dogma. Rote memorization and regurgitation of facts were what was expected, an approach Gurdon couldn't manage. At Eton, the elite boy's school that had served as an incubator for princes, dukes, and earls since the fifteenth century, Gurdon's instructors grew increasingly dismayed by the teenager's efforts to grow caterpillars. Finally, one of his tutors proclaimed him scientifically incompetent. "I believe Gurdon has ideas about becoming a scientist," the biology master wrote. "On present showing, this is quite ridiculous. If he can't learn simple biological facts, he would have no chance of doing the work of a specialist, and it would be a sheer waste of time both on his part and of those who would have to teach him." Out of a class of 250 biology students, Gurdon found himself dead last.

Chastened, he briefly considered a military commission before turning his attention to classics: ancient Greek and Latin. If he could not study insects, he would resign himself to a career studying ancient languages and cultures. But then came the first twist of fate that would reshape his path: he applied to the classics department at Oxford but was informed that he could matriculate only if he chose another major. Even this, his second-choice career path, was off limits! But the offer contained a small glimmer of hope: Might he consider applying to one of the scientific departments, which still had room? Thinking this might mean a return to his beloved insects, Gurdon eagerly revised his application and, with the help of a private tutor, managed to accomplish something he had failed to do before: memorize facts.

Gurdon passed the necessary entrance exams and a year later entered Oxford's zoology department as a science concentrator. But then, he encountered yet another hurdle. His plan all along had been to switch to the entomology department, with its focus on beetles, moths, bees, and ants. But to Gurdon's surprise, his application to transfer was rejected. The insect biologists, it turned out, didn't want him. Fortune had tossed him back and forth, encouraged hope and then blocked his path.

In retrospect, this was the best thing that could have happened to the young biologist, for it drove him to seek out Michail Fischberg, the friendly and supportive instructor who had taught his embryology class. The subject didn't captivate Gurdon the way insects had, but at least it held some appeal. Embryologists, he observed, paid great attention to the shape and form of growing embryos, and this reminded him of the complex wing and thorax patterns he had grown to cherish in caterpillars and butterflies. Gurdon knew little about embryonic development aside from what he had picked up in the class, but that didn't matter. Fischberg was ready to give him a chance when no one else would. At the age of 23, Gurdon claimed a desk in Fischberg's laboratory and began to ponder the mysteries of the embryo.

COUNTING GENES

A CENTURY EARLIER, ANOTHER INSECT LOVER—GERMAN BIOLOGIST August Weismann—had proposed his *germ plasm theory*, the model of development that had inspired Roux to perform his half-embryo experiments. The theory's cornerstone was the premise that each cell carried a distinct constellation of hereditary factors he called "determinants" (a conceptual precursor to genes) that were distributed unequally during cell division. Weismann postulated that a cell's place in the cellular society—whether a muscle cell, a blood cell, or a neuron—was dictated by the combination of determinants it carried. Accordingly, the zygote—as the single cell from which all other cells derive—would have to contain a complete set of determinants. The germ plasm theory thus viewed differentiation as a process of genetic dilution, in which a cell's place in the cellular society was determined by which genes it retained and which it lost during the process of division.

Weismann's theory can perhaps be best understood by seeing how it would apply to human, rather than cellular, career choices. Imagine that a baby is born with a comprehensive library of books, containing guides for every conceivable profession—volumes dealing with construction, farming, medicine, yoga instruction, and so forth. Now, picture an omniscient librarian who selectively removes book after book as the child ages, until only a few volumes remain. Scanning the half-empty bookshelf for guidance in choosing a direction, the young person, now having come of age, would see that their professional assignment had already been made, a decision defined by the collection of texts that were left on the shelf. In other words, the theory was a model of construction through loss, a partitioning of instructions whose directives were determined by whichever orders persisted at the end of development.

Weismann's theory also made an important prediction. If the model was correct—if the road of development was indeed littered with genes shed by cells as they specialized—it would mean that differentiation was a one-way street. Once settled into a particular profession, a cell would be unable to switch to another, for the guidebook that might allow for a career change simply wouldn't exist.

Once a muscle cell, always a muscle cell; once a neuron, always a neuron.

WEISMANN'S THEORY, PUBLISHED IN 1883, WAS ELEGANT BUT DIFFIcult to test. True, Hans Driesch had shown that the earliest cells of the sea urchin embryo were fully capable of giving rise to a new animal, a clear indication that they had not lost any such determinants. But the experiment said nothing about what might happen subsequently, as cells settled into their respective professions. Perhaps Weismann's determinants began the practice of segregation only at more advanced stages. Indeed, even Driesch had observed that a blastomere's ability to form an animal began to wane as the embryo grew older, as once the embryo had more than eight cells, their individual ability to form a new sea urchin was virtually nil.

Since the model predicted that each type of cell should carry a different and unique genetic complement, the most straightforward way to test Weismann's theory would have been to count the genes contained within diverse tissues. But until recently, there existed no way to conduct such a directed inventory. Even now, with sophisticated tools at our disposal, accurately counting genes in a single cell is nontrivial. Determining whether Weismann's theory applied to cellular differentiation required another way to peer inside the black box of development.

One possible solution came a half century later from Bob Briggs, a chromosome scientist with an infectious smile and a passion for research. Like Gurdon, Briggs had faced tough challenges on the road to becoming a scientist. Growing up poor, and raised by his grandparents in rural New Hampshire, he earned money by working at a shoe factory and playing the banjo for a small dance band. But once he realized that a career in science was a possibility, he pushed himself as hard as he could, earning a PhD from Harvard during the Great Depression. After graduating, he secured a laboratory position at the Lankenau Hospital Research Institute in Philadelphia, where he focused his attention on chromosomes.

Briggs's early experiments sought to understand the relationship between chromosome content and development. Theodor Boveri had addressed this type of question in the sea urchin, and Briggs wanted to see whether the mouse behaved similarly. His experiments created cells with too many or too few chromosomes, and in the process, he came up with an idea for an unusual experiment: What would happen if he took a cell's nucleus, with all its chromosomes and genetic material in tow, and transplanted it into an egg cell whose nucleus had been removed? Hans Spemann had proposed such a "fantastical experiment" years earlier, wondering whether a new animal would grow from the newly implanted nucleus. With his exceptional skill and experience, Spemann had come close to achieving this goal of *nuclear transplantation*, but never realized its full execution. Briggs, with more modern equipment, believed he could do it.

The reviewers at the National Institutes of Health, where he applied for a grant, rejected his proposal, calling it "a hare-brained scheme." But Briggs persevered, and with the funds he managed to scrape together he hired Thomas King, a graduate student at New York University with experience in microsurgery. King proposed using the northern leopard frog, *Rana pipiens*, whose large eggs would nicely suit their needs. Briggs agreed, and the work began.

King discovered that by puncturing the outer membrane of a frog egg with a micropipette (a thin glass filament that served as a microscopic syringe) and applying gentle suction, he could extract the cell's nucleus, resulting in an "enucleated" cell body—a cellular sack devoid of its genetic material. Then, using the same procedure, he was able to isolate nuclei from *Rana* cells at the blastula stage of development, which he then transferred into the enucleated cell body—thus replacing the nucleus that had once been there with a new one. With Briggs working as enucleator and King as injector, the duo refined this nuclear transplantation technique over the two-year period from 1950 to 1952. Remarkably, it worked. Not only did most cells survive the invasive procedure; at least a third of the hybrids grew and matured into tadpoles. Briggs's "hare-

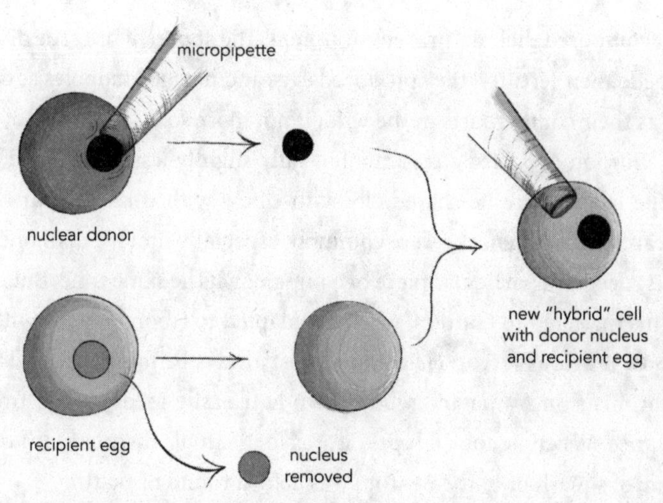

During nuclear transplantation, a micropipette is used to isolate the nucleus from one cell (the donor). In parallel, the nucleus of an egg is removed, leaving only an enucleated cell body (the host). When the donor nucleus is transplanted into the recipient egg, the resulting "hybrid cell" is allowed to develop. If such a hybrid can grow into a mature animal, the donor nucleus contains all the information (genes) necessary for development to proceed.

brained scheme" had tricked the nucleus into thinking it resided in a fertilized egg. Without realizing it, he and King had taken the first step toward cloning an animal.

THE FROG LAB

IN OXFORD, GURDON WAS SETTLING INTO LIFE AS AN EXPERIMENTAL embryologist. Fischberg's lab was a frog lab, which meant that space was cramped, and money was tight. The slippery amphibians occupied much of the laboratory's small footprint, as they were the ones who served as the raw material for the scientists' work. Their tanks—separate ones for housing and for breeding—were stacked from floor to ceiling, leaving limited space for the researchers. In Fischberg's lab the frogs lived out their lives in clean water, a change from the mud- and algae-ridden ponds that composed their natural environment. The apparent upgrade did not impede their fertility: they produced eggs and hatched tadpoles as capably as their counterparts in the wild, if not more so.

Gurdon occupied a separate but only slightly less constricted part of the lab, a space he shared elbow-to-elbow with three to four other researchers. Bottlenecks were common, especially if more than one scientist needed a particular piece of equipment at the same time. But none of this mattered to Gurdon, who had adapted to laboratory life with an almost religious fervor. He could almost always be found there, day or night, his thin frame and wild brown hair easily recognizable from a distance. When he took a break, it was for a simple distraction: a quick game of squash or, if the weather was cold, a round of skating.

What inspired Gurdon to work with such hunger? Was it to prove his worth to those who had rejected him in the past, like his science tutor at Eton or the entomology faculty at Oxford? Was he driven to reassure his parents, who had come to view their son as an academic wanderer, that he had now assuredly found his way? Or perhaps he was simply relieved that he had at last found someplace where he fit—an environment where his status was not immediately at risk—and his hard work would ensure that things stayed that way?

While all these factors may have contributed to Gurdon's drive, his dominant motivation, one shared by all serious scientists, was the hunt—the chance to work on a puzzle that was his alone to solve. To be the first person, ever, to lay eyes on a new fact of nature and give it a name or describe its features. Gurdon had grown up learning to appreciate the natural world on his own terms, rejecting the pedagogy of the classroom in favor of self-designed paths to knowledge, and this independence fueled his metamorphosis from neophyte to experimentalist. In Fischberg's lab, the fact that he was almost completely on his own was not a burden. It was a gift.

There was another element at play, a second childhood preoccupation that complemented his passion for insects: Gurdon had a fascination with scale models. Creating tiny objects with one's hands—miniaturized trains, boats, planes, and more—was a common enough pastime in twentieth-century England, but Gurdon took it to a higher level. His models—like the miniaturized British warship he constructed in the hollowed-out half of a walnut shell—were as detailed and demanding as anything an older and more experienced hobbyist could make. So, when Fischberg one day assigned his new graduate student the task of mastering nuclear transplantation, Briggs and King's technique that would require exquisite dexterity on a small scale, Gurdon couldn't believe his good fortune. It was the kind of thing he would have done for the pure fun of it, and now it would earn him a PhD. The task was an ideal marriage of hands and mind and would become the vessel into which Gurdon could pour all the painful frustrations and about-faces that had come before. After a long struggle to find his place in academia, at last he was home.

ONE DAY, AS GURDON REACHED INTO THE HOLDING TANK, THE FEMALE frog inside got startled. She arched her gray-green body and then kicked out her back legs with sufficient strength to splash him with water and disturb the numerous eggs resting on the bottom of the container. Gurdon had considered working with these eggs, which she laid the night before, but he knew from trial and error that things tended to go better if he harvested eggs that were fresh. So he moved quickly and decisively, grasping the frog's gelatin-covered skin firmly with two hands so she could not slip away. Then, he pressed his index finger against her cloaca, mimicking "amplexus"—the mounting posture of a male frog as he prepares to fertilize newly released eggs. After a few seconds, Gurdon eased the frog's legs apart and applied pressure to the base of her spine, setting off an involuntary impulse in some clandestine corner of her nervous system. At first, the frog fought, kicking in an attempt to break free of his grip. But as Gurdon continued to hold firm, she settled down to perform the job that nature had assigned her millions of years earlier and began to lay new eggs by the hundreds, which he collected in a container positioned underneath.

They were beautiful spheres, darkly pigmented on one side and pale yellow on the other, like obliquely illuminated celestial bodies. Gurdon placed them in a glass petri dish, and with an eyedropper he blanketed them with a small amount of frog sperm. By inseminating the eggs in this way, rather than allowing male and female frogs to mate, he ensured that all would be fertilized at the same time. Within minutes, most of the eggs were penetrated by the sperm. Then, the newly formed embryos began their march to maturity, doubling with precision every 20 minutes. Like those of all animal species, these embryos didn't grow larger with each round of cleavage; rather, each of the daughter cells became half as big as its parent. Within hours, the petri dish was teeming with dozens of blastulae, each no bigger than the eggs from which they were derived despite the hundreds of cells packed tightly within. Within days, swimming tadpoles filled the space.

As he watched this developmental procession, Gurdon wondered what happened when one of the primitive cells of a blastula took on more of the specialized attributes of a tadpole or, later, a frog. The tadpole needed a tail to propel it toward nutrients or away from predators, an intestinal tract to digest food, gills to breathe, a nervous system to control it all, and so on. But all these components were lacking at the initial stages of development. Were genes and chromosomes getting dismantled and sorted into different cells, as Weismann's germ plasm theory predicted? Was this the secret behind differentiation? He didn't know, but he began to believe that nuclear transplantation—the technique he had been assigned to master—could be a way to find out.

GURDON'S FIRST ATTEMPTS WERE FAILURES. NO SCIENTIST AT OXFORD (and few anywhere in the world) had experience with Briggs and King's procedure. The technical descriptions in their paper were accurate, but a written account could never capture the nuances of the method, leaving Gurdon with only a rough map to go by. His first challenge, making the micropipettes, had come relatively easily. To fashion the fine syringes, Gurdon obtained hollow glass rods—each the size of a pencil lead—and suspended them over a flame until their centers became soft. He then pulled smartly on each end, stretching the glass tubes until their centers were thinner than the diameter of a cell; with practice, he could make a dozen of these microsyringes in an hour, more than enough for a day's worth of experiments.

The big problem was the microinjection itself. When Gurdon brought the fine tip of his mini-harpoon up against the egg's outer membrane and pushed, nothing happened. Instead of puncturing the soft shell, as he intended, the micropipette merely deformed it, just as a finger pushed into a water-filled balloon creates a depression but fails to puncture its outer lining. The gelatinous skin surrounding the eggs was too elastic, and there was no way to harvest the precious nucleus inside. Briggs and King hadn't faced this difficulty, as the membrane covering the eggs of *Rana pipiens*, the frog species they used, was rigid, allowing their needles to penetrate easily. But Fischberg's lab used the African clawed frog, *Xenopus laevis*, which was now the root of the problem. Gurdon adjusted his technique, making even finer pipettes in the hope that a sharper point would get through more easily, but this only caused the pipette's tip to snap off from the pressure. He was at an impasse.

Fate had previously intervened on Gurdon's behalf at critical moments, and it now came to his rescue again, this time in the form of a microscope. While most microscopes relied on simple light—a plain bulb—to illuminate their subjects, Fischberg had recently purchased a new microscope that used an ultraviolet (UV) light source. By chance, Gurdon discovered that shining the UV light on the eggs loosened their jelly, thus finally granting him access to the cell's interior. But Gurdon's good fortune didn't end there. As he then began poking his pipettes into eggs in preparation for plucking out their nuclei, he found that the UV had destroyed the host-cell nucleus along with its genetic material. It was a two-for-one bargain: the same exposure had given him access to the cell and simultaneously destroyed the compartment he needed to extract.

Suddenly, life was becoming easier. Now, simply exposing the eggs to UV light would condition them as nuclear recipients, rendering enucleation unnecessary. Harvesting the donor nuclei turned out to be even more straightforward, as these would come from more mature cells—at the blastula stage or later—that were easily penetrated by Gurdon's micropipettes without UV exposure.

With all elements of the procedure in hand, Gurdon had finally hit his stride. Over the next few months, he mastered the transplantation technique. To his satisfaction, he easily replicated Briggs and King's results, coaxing nuclei extracted from blastulae to turn into tadpoles and even frogs by transplanting them into irradiated eggs. Not satisfied with this simple success, he continued to experiment and improve until each step could be performed with maximal efficiency. Now, he could perform dozens of transplantations in a single sitting. Nuclear transplantation had become as natural to him as driving a car or playing squash.

RESETTING THE DEVELOPMENTAL CLOCK

MEANWHILE, IN PHILADELPHIA, BRIGGS AND KING HAD BEEN HARD at work extending their results. Their earlier success—generating new organisms from transplanted nuclei—was a technical tour de force that had secured their reputations. But their procedure had been just that—a procedure—rather than a real conceptual or practical breakthrough.

By then, however, Briggs had realized that the procedure could serve as a testing ground for the germ plasm theory. If Weismann was right, it followed that more specialized cells should contain fewer determinants (genes) than undifferentiated ones. Consequently, the success of nuclear transplantation indicated that the nuclei they had used in their initial experiments hadn't lost a significant number of genes; otherwise, how could they have given rise to whole new frogs?

But so far, Briggs and King had used only nuclei taken from cells of the blastula, a stage of development that preceded most differentiation. In other words, they had been looking too early to say anything about Weismann's model. If they really wanted to put the germ plasm theory to the test, the donor nuclei would need to come from more specialized cells—cells that had begun to take on the features of different organs. Only then could they ask the critical question: Would the transferred nuclei still support development? In effect, nuclear transplantation could be a way to turn the gene-counting problem on its head—instead of trying to measure whether cells lost genes as they specialized, the technique could interrogate cells *after* differentiation to ask whether something had been lost along the way.

Briggs and King thus turned their attention to more mature embryos—those that had begun the process of specialization ("gastrula" stage) or had gone so far as to form a primitive nervous system ("neurula" stage) or the beginnings of a tail (tadpole stage). Extracting nuclei from various parts of these embryos, the duo assessed the success of nuclear transplantation at each phase of development. Now, the odds that their hybrid embryos would develop dropped precipitously with each step in the differentiation process. Nuclei from the more advanced gastrulae performed half as well as those from the more immature blastulae, while those taken from the neurula stage had barely any developmental potential. By the time an embryo had started to develop a tailbud or a heartbeat, the odds of success dropped to zero.

The germ plasm theory had predicted just such a result. If becoming a specialized cell—a neuron, say—meant losing the instructions for any other type of cells, then the nucleus of such a cell would no longer be competent to support development. Indeed, the same limitation should hold true for all differentiated cells. On face value, these new experiments supported Weismann's model of gene forfeiture.

DESPITE (OR PERHAPS BECAUSE OF) HIS INEXPERIENCE, GURDON HAD a problem with this interpretation. Briggs and King had obtained what is referred to as a "negative result," or what might more aptly be called a "nonresult"—the failure of an experiment to yield a particular sought-after outcome. When this happens, it leaves a scientist in the tricky position of deciding whether the lack of findings is meaningful or not. Instead of growing into tadpoles, as their previous nuclear transplants had done, Briggs and King's new transplants resulted in *a dearth of tadpoles*—a result that could have been due to any number of confounding factors. For example, if specialized cells happened to be more fragile than their immature counterparts, their nuclei might have been prone to damage during the delicate transplantation procedure. If so, a technical issue, rather than a loss of genes, might have been responsible for the lack of success.

In Gurdon's view, Briggs and King's results neither confirmed nor refuted Weismann's theory. He had found their initial studies convincing (he had replicated their results, after all), but he considered their subsequent conclusions to be on shakier ground. He knew firsthand how difficult nuclear transplantation could be, and so it was easy for him to imagine how a technical glitch, rather than biology, could have led the Philadelphia scientists to a mistaken conclusion.

But Gurdon was taking a professional risk by questioning their interpretation. By the late 1950s, Briggs had earned an exalted position in the scientific community as the inventor of nuclear transplantation, while Gurdon was still cutting his teeth in Fischberg's laboratory. If he ever hoped to graduate, he would need his work to be published in a scientific journal, and challenging the established dogma was clearly not the best strategy. Gurdon's colleagues cautioned him to find another project. What made him think he could compete against the inventors of the very technique he was now using?

It was sensible advice, but all the difficulties he had overcome to get to this point had made him bold. Young, stubborn, and not yet possessing a reputation that could be seriously damaged, Gurdon pressed on, convinced that if he remained focused and performed the right experiments everything would work out.

IN HIS ATTEMPTS TO MASTER NUCLEAR TRANSPLANTATION, GURDON had made two changes to Briggs and King's original protocol. The first was the manner of enucleation—the Americans had removed the egg's nucleus before transplantation, while Gurdon used ultraviolet light to obliterate the egg's genetic material. The second and more critical difference involved the experimental model. *Rana pipiens*, the northern leopard frog species that Briggs and King had employed, may have had large eggs and large nuclei that facilitated their manipulations, but the species also presented one major drawback. Females produced eggs only in the spring, which had forced the Americans to complete all their experiments in brief windows of time.

This, it turned out, had been the basis for Fischberg's decision to select *Xenopus laevis* as his laboratory's experimental animal of choice. Unlike with *Rana*, researchers could induce *Xenopus* females to lay eggs by injecting them with fertility hormones, meaning that they could obtain embryos throughout the year. In retrospect, the use of *Xenopus* eggs, whose gelatinous outer skin had given Gurdon such trouble initially, now gave him a huge advantage, for he could do his experiments whenever he wished. So Gurdon was now in complete control, doing dozens of transplants in a day, his hands performing the delicate procedure by instinct. His goal had changed from replicating old findings to making new discoveries, as he wondered whether he might see something different than Briggs and King had. He was ready, finally, to bring all his technical and intellectual expertise to bear on a single (and seemingly simple) question: Do differentiated cells retain all the instructions needed to create a new organism?

It did not take long to get an answer. In lieu of the immature embryonic cells he had earlier used, Gurdon now extracted nuclei from cells that had already undergone differentiation, in some cases using cells more developmentally advanced than the ones Briggs and King had used. If the conclusions of the two Philadelphia scientists were correct, these cells should have lost genes prescribing some alternate fate—a genetic molting of sorts. Such genes would certainly be necessary to make a frog from one cell, and thus Gurdon's new round of experiments should not have worked, as Briggs and King's had not.

Except that they did.

When Gurdon made hybrids with these more mature embryonic nuclei—embryos that had a heartbeat or a nervous system—these too were able to undergo normal development, giving rise to normal tadpoles and, in some cases, even frogs. Nuclear transplantation appeared to be a kind of cellular baptism by which old cells could be born again. Gurdon's transplants didn't work all the time—indeed, the chances of success decreased as more mature nuclei were used, just as Briggs and King had observed. But the fact that the experiment worked at all was astonishing. This was no negative result! Nuclei that only a day before had been performing highly specialized jobs were now directing the formation of new living creatures. As he would later reminisce, Gurdon was watching differentiated cells forget everything they had previously learned and begin life anew.

AN ABUNDANCE OF CAUTION

SCIENTISTS ARE NOT, AS A RULE, A CELEBRATORY BUNCH. REVELRY IS the exception, not the rule, in the laboratory, as the exhilaration of discovery soon gives way to doubt. What mistakes could I have made along the way? What confounders did I fail to consider, or what miscalculations might have sent me down a fruitless path? These questions become especially pressing when one's results conflict with previously reported findings, and doubly so if those findings are from a well-respected research group.

After this gauntlet of self-criticism has been crossed, there is yet another hurdle: the peer-review process. Scientific peer review—the skepticism-laden practice of having others in the field scrutinize research findings prior to publication—is designed to identify the technical errors, missing controls, or flawed logic you didn't consider. As a result, elation over even the most significant findings tends to be short-lived; by the time a result has been thoroughly evaluated, criticized, taken apart, put back together, and ultimately accepted for publication, the excitement of discovery is long gone.

What should have been a thrilling moment for Gurdon, a time for him to rejoice in his discovery, ended up ephemeral and subdued. As he watched tadpoles and frogs emerging from his latest set of transplants, his first thought was "This is amazing." But almost immediately, apprehension set in. "Could it actually be right?" "Could there be an artifact," the result of some technical error or miscalculation? Gurdon knew he faced an uphill battle, that others would surely be dubious or dismissive. He was a newbie, after all, an apprentice who had only started in the lab a year earlier. His observations directly contradicted Briggs and King, who had pioneered the very method he was using. Why should the world accept his results over theirs? Even Fischberg—as supportive as he was—secretly believed that his student must have made a mistake somewhere along the line.

Gurdon knew that most of the criticism would be directed at his methods. Critics would take issue with the fact that he had used UV illumination to disable the host cell's nucleus, thereby skipping the enucleation step. This had made the procedure easier, but it also raised an unsettling prospect: What if Gurdon only *thought* he had destroyed the egg's nucleus, but in reality, he hadn't? Even if the nucleus appeared to have shattered, perhaps some or all of its genes lived on. If this were true, the hybrids might have two potential sources of genetic information—genes from the transplanted (donor) nucleus and genes from the host-egg nucleus that had never been destroyed. From the experiments he had done so far, there would be no way to know whether either or both had contributed to the resulting tadpoles and frogs. He could imagine the scorn he would face from critics—appraisals that would be as demoralizing as the evaluation from his long-ago tutor proclaiming him scientifically incompetent.

"You have done nothing besides showing that a frog egg can give rise to a frog!" they would surely say. "If you're so sure that ultraviolet light destroyed the host nucleus, *prove it*."

HOW TO PROCEED?

Gurdon considered revising the protocol but quickly dismissed this approach. The UV shortcut (which he still believed worked as intended) had been a gift, and he was loath to give it up. Extracting and discarding the host-cell nucleus, as Briggs and King had done, would have been the more definitive approach, but it would mean going back to square one.

Fischberg came to the rescue. Gurdon's supervisor reasoned that his student simply needed a way to distinguish between donor and host cells, the trick Spemann and Mangold had used in their organizer experiments three decades earlier. Having such a "genetic marker" would make it obvious which nucleus—donor or recipient—had been the source of the new animal. Better yet, Fischberg had the perfect tool in mind. It was another strain of *Xenopus*, identical to the one Gurdon was using in all respects save one: its cells had a defect in their "nucleoli," small structures within the nucleus, that made them easily distinguishable under the microscope. This marker would thus indicate whether a frog was the product of the transplanted nucleus, with its distinctive nucleoli, or the host nucleus, with its normal nucleoli. One can only imagine Gurdon's relief when the results came out as he had anticipated. The tadpoles and frogs arising from his hybrid eggs invariably bore the mark of the donor. The nucleus, that subcellular carrier of genetic information, could spawn a new animal, complete in all its parts.

Gurdon thought of other ways to strengthen his case. Instead of choosing randomly which cells would serve as transplantation donors, he selected cells he was sure had differentiated based on their appearance. It would be difficult for his critics to argue that such donors—like certain intestinal cells whose fingerlike projections (called "microvilli") carry out specialized absorptive functions—were not already differentiated. Seeking further proof that the transplanted nuclei hadn't lost any genes, Gurdon showed that nuclei taken from an animal produced by nuclear transplantation could also seed other generations with the same efficiency. The process of nuclear transplantation could be repeated again and again, giving rise to multiple generations of offspring.

Over time, Gurdon found even better markers that could distin-

To prove that new frogs arose from the nucleus of the donor rather than the host, Gurdon transplanted nuclei taken from an albino (white) frog into eggs from a normal-pigmented female (middle, top). The donor-derived offspring were all albino—a clutch of clones!

guish donor from host. One of the most dramatic demonstrations made use of nuclei taken from white albino frogs, which, when transplanted into the eggs of normal pigmented females, gave rise to albino offspring. As his technique improved, Gurdon's success extended to more mature nuclear donors—cells whose specialized features were unambiguous. It was an onslaught of data, and even Briggs, whose results had been thus challenged, came to accept the Oxford scientist's findings. By resetting the developmental clock of even the most specialized cells, Gurdon had defined a fundamental feature of development. The age of cloning had begun.

GENOMIC EQUIVALENCE

IN THE NOVEL *BRAVE NEW WORLD*, ALDOUS HUXLEY GAVE LIFE TO A fictional civilization in which human cloning was routine. In one of the opening scenes, he describes a tour of the factory in which the so-called Bokanovsky process is used to generate large numbers of identical individuals:

> "Bokanovsky's Process," repeated the Director, and the students underlined the words in their little notebooks. One egg, one embryo, one adult—normality. But a bokanovskified egg will bud, will proliferate, will divide. From eight to ninety-six buds, and every bud will grow into a perfectly formed embryo, and every embryo into a full-sized adult. Making ninety-six human beings grow where only one grew before. Progress.

In pioneering the technique of nuclear transplantation in *Rana pipiens*, Briggs and King had taken the first steps toward cloning an animal (they even used this term in some of their initial studies). Gurdon had taken the method to a different level by creating animals from the nuclei of fully differentiated cells. His work suggested that it might be feasible, at least in principle, to make a new and genetically identical creature—a clone—from a nucleus obtained from any animal species, a prospect that could include human beings. But aside from the social and ethical implications of this, which we'll return to later, Gurdon's result held a radical implication for biology. The fact that a nucleus taken from a differentiated cell could give rise to an animal meant that the genetic instructions needed for development still resided within that nucleus. Weismann was finally proved wrong—cells do not lose genes as they differentiate. Instead, all cells in the body carry a complete set of genes, a principle known as *genomic equivalence.*

In the wake of Gurdon's findings, scores of scientists around the world attempted to extend his results to other species, particularly mammals. They tried replicating Gurdon's success by using rabbits, mice, pigs, cows, and monkeys, but these efforts failed. While the procedure was occasionally successful using nuclei extracted at the very earliest stages of embryogenesis—when embryos consisted of fewer than 8 or 16 cells—no one could get nuclear transplantation to work with nuclei from more mature mammalian embryos, much less adult animals. Scientists began to wonder whether frogs were alone in their ability to be cloned. Was genomic equivalence a unique property of these amphibious creatures?

In 1997, nearly 40 years after Gurdon's discovery, these concerns were put to rest. A Scottish team led by Keith Campbell and Ian Wilmut created Dolly, a sheep made by transplanting the nucleus from an adult ovine mammary gland into an enucleated sheep egg. Just as the switch from *Rana* to *Xenopus* had been critical for Gurdon's success, it was a small technical detail—a feature of the internal clocks of cells—that made all the difference for Campbell and Wilmut. Eggs normally exist in a dormant state, resting unobtrusively for months or years until fertilization shakes them out of their slumber to divide. But in their failed attempts at mammalian nuclear transplantation in the 1960s, '70s, and '80s, most scientists had neglected to take this into account, using laboratory-cultured cells that were dividing rapidly. Campbell and Wilmut hypothesized that the egg, in its arrested state of division, might be unprepared for a nucleus derived from a rapidly dividing cell—a mismatch that could have doomed the cloning process. Campbell and Wilmut thus tried a simple fix: slow down the rate of division before extracting the nucleus for transplantation, in effect lulling the donor cell back to sleep. It worked, and Dolly was the result.

Nuclear transplantation is technically challenging in mammals, with success rates much lower than those in frogs. Even so, Campbell and Wilmut's success motivated an explosion of mammalian cloning. To date, nuclear transplantation has been performed successfully thousands of times, resulting in a clone menagerie that includes mice, rats, cats, dogs, goats, water buffalo, mules, horses, gaur, camels, and monkeys. Clones have even been generated years after an animal has died, using nuclei derived from its frozen body. With each success, the case for genomic equivalence has become stronger, and its dogma is this:

Cells hold on to all their genes—even those they don't need—from zygote to adulthood.

Despite claims to the contrary, there have been no credible reports of cloned human babies. But if history is any indication, any technical issues standing in the way are surmountable. As we will see later, the ethical considerations involved in this type of reproductive technology are complex. For example, the prospect of creating genetically identical human beings—Bokanovsky's process—is frightening. But what about the more heartening prospect of using nuclear transplantation to create healthy cells to replace sick ones—an advance that could result in treatments for various degenerative diseases? Regardless of where cloning technology leads, it opens the door to an array of possibilities that Briggs, King, and Gurdon could not have imagined when they started to poke frog eggs with tiny needles.

GURDON WOULD BE THE FIRST TO POINT OUT THE OUTSIZE ROLE SERendipity played in his career, from the twists of fate that diverted him away from and then back toward science when he was young, to the happenstance procedural modifications that allowed him, as a student, to detect what others had missed. In many ways, his path echoed the unpredictable journey that cells take as they find their own way in the cellular society, following their internal compasses at the same time that they respond to outside influences. For as we have already seen, embryonic development is the archetypal example of nature and nurture working in synergy.

Gurdon, who has won nearly every biology award there is (including the 2012 Nobel Prize in Medicine), came to view the events that sidetracked him as a young man—the discouraging letter from his tutor, his inability to get a military commission or a job, and his rejection by the Department of Entomology—as a series of fortunate accidents. For in drawing him away from insects, these chance events led him to make contributions that would have been impossible had he succeeded in fashioning the career he had initially imagined for himself.

Nevertheless, Gurdon still gets a look of delight in his eye when the subject turns to the flying creatures that got him hooked on biology. "Someday, if I don't have a laboratory, I can think of some experiments I'd like to do with insects, to figure out what gives moths their patterns," he muses, recalling how church vicars in the eighteenth and nineteenth centuries spent years cataloging moth and butterfly patterns in excruciating detail. They are classifications that modern entomologists still rely upon, their descriptions so comprehensive that they provide insight into not only the decorations that nature created but also the ones it chose to omit. Using his hands animatedly to illustrate the point, Gurdon concedes that the beautiful patterns that decorate the bodies of these creatures may be too complex to be understood in simple genetic terms.

In some ways, this is the essence of what Gurdon's experiments taught us: that genes alone do not dictate fate. Despite their extraordinary diversity, all cells in the body carry the same genetic information—proof that genetic content alone is insufficient to stipulate a cell's destiny. A neuron "knows" that it is a neuron, and a muscle cell "knows" that it is a muscle cell, even though both cells carry the same genetic instructions.

This simple fact creates a crucial paradox, for if the presence or absence of a gene doesn't determine a cell's fate, what does? How does nature coordinate the position, shape, and function of the hundreds, millions, or trillions of cells in an animal's body if every one of them has the exact same genes? The answers to these questions have come from an unlikely source: studies of the eating habits of bacteria and the sleeping habits of viruses.

Chapter 4

TURNING GENES ON AND OFF

Science says the first word on everything
and the last word on nothing.

— VICTOR HUGO, *INTELLECTUAL AUTOBIOGRAPHY*

What Eliot said of poetry is also true of DNA:
"All meanings depend on the key of interpretation."

— JONAH LEHRER, *PROUST WAS A NEUROSCIENTIST*

François Jacob had just turned 30 when he walked into André Lwoff's office for what he feared would be the last time. Several years had passed since Jacob completed his military service, fighting the Nazis with the Free French Army in Tunisia and Normandy, and in that postwar interlude, he had lost all sense of personal direction. As a teenager, he had dreamed of becoming a surgeon, healing the sick with scalpel and bandages, but battlefield injuries to his legs—and the ensuing years of operations, complications, and rehabilitations—had left his body and spirit bruised. Returning to Paris, he completed his medical studies, knowing inside that he would never practice medicine. He tried his hand at odd jobs and other pursuits, but nothing suited him. Like so many others in that postwar metropolis, François Jacob became a man in search of a purpose, carried by whatever current of fate caught him on any given day.

Then, a dinner party and a chance introduction. One of the other guests—a cousin by marriage and, like Jacob, a 30-year-old veteran—had a story with an uncanny resemblance to his own. Herbert, as the man was named, had also aspired to a medical career, only to abandon that ambition after the war. But there the similarity ended, for Herbert had done something radical: without any prior qualifications, he embarked on a career in science. Jacob was taken aback. Such reinvention—picking up a new and intellectually challenging business—had never even crossed his mind. What audacity to set one's sights on such an improbable goal (especially at the advanced age of 30), and then to actually make it happen.

As he listened to his new acquaintance describe the daily routine of the laboratory, where his job was to maintain cultures of yeast and record their properties, Jacob could discern nothing about Herbert that would seem to make him uniquely suited to the profession. He seemed no smarter or more knowledgeable than Jacob, nor did he possess any unique skills or talents before making his career transition. And yet, with no experience, this man had calmly found his way into an established and respected Parisian laboratory.

Jacob could feel an emotion welling up inside him—one he had not experienced in a decade: hope.

With fresh determination, Jacob applied to study at the Institut Pasteur—the world-famous research center named for Louis Pasteur, the researcher who a century earlier had proved that microbes cause disease. Jacob was admitted and took the institute's "Grand Cours," a yearlong survey of bacteriology, virology, and immunology. The rigorous methods and simple truths of science resonated with him, and at the end of the year it was time to find a laboratory mentor, someone who would teach him the practical aspects of research.

Jacob approached André Lwoff, the institute's tall and refined director of microbiology. Lwoff's lab studied bacteriophage, the bacteria-seeking viruses that Alfred Hershey and Martha Chase used to confirm

DNA's genetic properties. But Jacob was drawn more by Lwoff's reputation than the specifics of his research—Lwoff, he believed, would make a real scientist out of him. But when Jacob called on the senior researcher, repeatedly, to ask for a laboratory position, he was rebuffed. On each occasion, Lwoff received the novice with courtesy, then politely told him that the laboratory was full. (The truth was that Jacob lacked experience, and Lwoff could afford to be picky.)

But one day in the summer of 1950, when Jacob had nearly abandoned hope of joining Lwoff's lab, there was something different in the director's manner—his mood was friendlier, and his eyes betrayed an inner warmth.

"You know, we have just found the induction of the *prophage*," Lwoff announced with a smile.

"No, really!?" Jacob replied with as much incredulity and excitement as he could muster. He had never heard of a prophage or its induction, nor could he find the terms in the dictionary later that night. But seven years of wandering had given him a boldness he didn't realize he possessed until that moment.

"Would *you* be interested in working on the phage?" Lwoff prodded.

"That's just what I would like to do," Jacob answered, leaping headfirst into his future.

THE PROPHAGE LWOFF SPOKE OF WITH SUCH ENTHUSIASM IS A satellite-shaped virus named lambda (abbreviated with the Greek symbol λ). Like all viruses, λ is incapable of reproducing on its own and relies on its host, the gut bacterium *Escherichia coli*, to manufacture new viral particles. The virus first latches on to a receptor on the surface of the bacterial cell—a protein the virus has evolved to use as a docking station—and then introduces its DNA into the bacterial cell.

With its genome safely inside the host, the λ phage must make a simple yet critical decision: whether to remain awake or go to sleep. To a large extent, this decision depends on the metabolic state of the cell at the time of infection. If adequate nutrients are available, the virus takes advantage of the opportunity and reproduces. It accomplishes this feat by coaxing the cell's DNA-replication machinery into action, generating dozens of new viral particles. Eventually, the burden becomes too great for the host, and the cell bursts open, or "lyses," under the strain. This replicative state is thus called the *lytic cycle*. If, on the other hand, nutrients are scarce, the virus enters a state of dormancy and hibernates until conditions are more promising. This dormant state is called the *lysogenic cycle*.

The consequences of this choice—lysis versus lysogeny—are significant for both phage and host. Lysogeny allows the bacterial cell to go about its business as the invader bides its time, while lysis means death for the bacterium and new life for the virus. When the phage assumes a lysogenic state, slumbering in the comfort of a bacterial chromosome, it is called a prophage. So when Lwoff announced to Jacob that he had discovered how to induce the prophage—which could be achieved by simply exposing cells to ultraviolet light—he was describing a means of rousing the phage from its rest, a kind of viral alarm clock.

Lwoff and Jacob could have no way of knowing that the systems regulating the behavior of these obscure phage would have direct relevance to the One Cell Problem. On the surface, these simple organisms bore little similarity to the far more intricate workings of the human body. But nature is thrifty, using the same solution to an evolutionary problem over and over, connecting microbes to humans in the process. Consequently, the assignment of a cell to its place in the cellular society bears an unmistakable resemblance to the forces governing the sleeping habits of phage.

THE ATTIC

EMBRYONIC DEVELOPMENT WOULD HAVE BEEN MUCH EASIER TO understand if August Weismann had been right—if a cell's fate were dictated by its genetic makeup. Changes in the genome are precisely how new species arise during evolution. But as John Gurdon showed us in the last chapter, a cell's path of differentiation is determined not by its genetic complement but by something else. Cells in the brain, kidney, heart, and lung bear little resemblance to one another, yet they all carry nearly identical DNA sequences. This phenomenon of genomic equivalence raises an obvious question: If every cell in the body carries a uniform and unvarying set of genes, why aren't all cells alike? What distinguishes one cell type from another?

The answer is *gene regulation*—a process that renders each segment of the genome either decipherable or concealed in a given cell.

Imagine embryonic development as a play, with genome as script and cells as cast and crew. Even before production starts, everyone involved—actors, production staff, designers, directors—receives a full copy of the script so that they can follow what is supposed to happen on stage from one moment to the next. It is not enough for an actor to know her own lines. She must know everything, because her character's motives, desires, fears, and abilities are all embedded in the script. The production crew must also have the entire script, as it provides the context for costumes, set designs, lighting, and sound. The script defines the play's arc—where the story starts and where it intends to go.

Likewise, developing cells each receive a full copy of the script—the genome—which they carry with them throughout life. As each cell learns its role, it highlights certain parts of the genetic text, making notes in the margins as an actor might do while learning their part. This is the essence of gene regulation, a marking up of the genome by which cells learn their lines—which ones to focus on and which ones to ignore. Progressively, through the rehearsals that continue throughout embryonic development, the important lines are reinforced while those of other players are repressed. By the time the animal is born, each cell has learned its role, identifiable on the body stage by the genes it expresses. Gene regulation gives each cell its voice, distinguishing it from every other cell in the body.

If you had asked André Lwoff to speculate how the phage switches between lysis and lysogeny, gene regulation would have been the last thing he said; at the time, the notion that the genetic material could be regulated was a completely foreign one. This was not because he doubted the importance of genes in biology. On the contrary, genes were viewed as so important that it was impossible to imagine their involvement in the more routine and mundane day-to-day activities of a cell. It was assumed that enzymes—proteins that facilitate the chemical reactions that occur within our cells—were solely responsible for a cell's conduct, a "mass action" model in which genes had no direct role. DNA, the genetic material, was assumed to act passively, responsible for initiating the marvel of development but completely divorced from its administration.

IN SEPTEMBER 1950, JACOB PRESENTED HIMSELF TO LWOFF'S LABORAtory on the top floor of the Pasteur Institute—a space beneath the building's majestic mansard roofs that Lwoff had nicknamed "le Grenier" (the Attic). The Attic had much in common with the Stazione Zoologica in Naples—that hotbed of scientific activity that had served as a scientific proving ground for Driesch, Boveri, and Morgan a half century earlier. Jacob was enthralled by the intensity and sense of purpose that permeated the Attic's cramped laboratories, where the lively intellectual life created its own expansive space—a realm of curiosity and discourse.

The Attic's hallways served as a center of activity—common areas where scientists could emerge from their laboratory rooms throughout the day to share their latest findings. But the other nexus for scientists to gather, it turned out, was Jacob's room, for it was the only place in the Attic with a table large enough to seat everyone. At one o'clock each day, the Attic's scientists appeared there with their carafes and lunchboxes filled with meat and cheese sandwiches and claimed one of the stools surrounding the table.

Then, the discussions would begin. Someone would propose an idea and then the group would take it apart, examine each piece in detail, and either reassemble it in a more fitting shape or discard it entirely. Every concept was snapped up by someone else, kneaded, turned over, ground, passed through a sieve, and finely pulverized before the group moved on to the next subject. The topic could be scientific, personal, or political, and could change in an instant, switching from serious to frivolous or from a group discussion to a one-on-one debate. Journal articles, recently released books, travel stories, war recollections, the role and responsibility of the scientists behind the atomic bomb, the insidious impact of McCarthyism, French politics—these were common themes of lunch in the Attic. But by the time two o'clock rolled around, Jacob, the former resistance fighter, could think of only one thing—the moment that everyone would clear out so he could continue the day's experiments.

EMBRYONIC DEVELOPMENT IS, BY DEFINITION, A FEATURE OF MULTI-cellular animals. But in the 1940s, biologists began heading in the opposite direction, moving away from the sea urchins, frogs, newts, and flies of their predecessors in favor of even simpler organisms—bacteria and their parasitic phage. Leading this shift was Max Delbrück, a Caltech physicist turned biologist who argued that it would be impossible to understand the biology of complex animals if one could not first understand the inner workings of the most basic unit of biology, the cell. Which cell and which organism, he argued, shouldn't matter. Indeed, the simpler the system, the better. Animals, Delbrück believed, were simply too messy to yield anything of fundamental importance.

There were also practical reasons driving the switch to microbes. Growing animal cells in culture was, and still can be, cumbersome and expensive, requiring defined culture conditions and aseptic technique (to keep the microbes out!). Bacterial cells, by contrast, grow in simple broths, needing only basic building blocks such as sugars, amino acids, and salts. When supplied with these materials, bacterial cells divide at astoundingly high rates—as frequently as once every 20 minutes. This means that a single bacterial cell can give rise to a thousand trillion progeny in just one day (even more, when it comes to phage).

Delbrück's vision propelled the molecular biologists of the mid-twentieth century, who amassed an impressive track record using microbes. Oswald Avery's conclusion that DNA is the genetic material (the "transforming principle") arose from experiments with the lethal pneumococcus bacterium, findings that were confirmed by Martha Chase and Alfred Hershey in their phage experiments. Delbrück himself used these microorganisms to show (with biologist Salvador Luria) that mutations arise spontaneously and not as the result of selective pressures from the environment, experimental proof for the principle of natural selection that Darwin had proposed almost a century earlier.

Yet despite the simplicity of the phage system, studying the induction of λ—what caused it to awaken in its bacterial host rather than remain dormant—turned out to be more complicated than either Lwoff or Jacob anticipated. There were many questions, but the tools available to address them were primitive. One persistent mystery had been the host's role in phage induction. It seemed unlikely that the phage had anything resembling free will, deciding on its own whether to sleep or stay awake. A more plausible scenario was that the bacterial cell had a say in the matter, and that some sort of interaction between virus and host was mediating this "decision." Jacob decided to devote his energies to the *E. coli* host.

After completing his PhD in 1954—a thesis that introduced him to the ways of microbiology but yielded few novel insights—Jacob befriended Élie Wollman, a fellow Pasteur researcher who was studying a form of bacterial mating known as *conjugation*. Discovered in the 1940s by Joshua Lederberg and Edward Tatum, conjugation allows cells to share genetic material as a means of coping with environmental stresses. (This type of genetic sharing is the primary mechanism by which antibiotic resistance spreads.) In the jargon of the field, a bacterial strain capable of passing genes along is called "male," while a bacterial strain that can receive but not transmit genetic material is called "female." The process was poorly understood, but Wollman, who had studied with Delbrück at Caltech, was making headway.

One of Wollman's insights was that conjugation could be used to create maps of bacterial chromosomes—similar to the maps of *Drosophila* chromosomes that Morgan and Sturtevant had made decades earlier. The method involved interrupting the microbial mating process—or the "conjugal bliss," as Wollman called it—at various time points: 10 minutes, 20 minutes, or 30 minutes after mixing male and female cells. By simply observing which traits were transferred during each interval, one could determine the chromosomal order of the genes encoding those traits. For example, if trait X was transferred after 10 minutes, traits X and Y were both transferred after 20 minutes, and traits X, Y, and Z were all transferred after 30 minutes, then the order of those genes on the chromosome must be X-Y-Z. The instrument for achieving the bacterial coitus interruptus came from Wollman in the form of a kitchen blender he had bought for Odile, his wife.

Jacob allied himself with Wollman, feeling his way through the endless calculations needed to generate chromosome-scale maps. In the process, he managed to squeeze in a few experiments of his own, studies that might shed light on phage induction. One result stood out. When Jacob conjugated cells carrying the sleeping phage (lysogenic males) to "naïve" cells (uninfected females), the virus awoke almost immediately. Jacob christened the phenomenon "zygotic induction." It was a new way to awaken the sleeping virus, one distinct from the method of UV exposure that Lwoff had been so excited about. But the result also held a deeper lesson, for it showed that it was not an alarm clock that caused the prophage to wake up. Rather, it was respite from a molecular sedative—an *inhibitor* of viral replication—that had roused the prophage. Conjugation had allowed the viral genome to enter a cell lacking such a sedative, and this had been enough to revive it. Phage do not sleep because they are *missing* something, Jacob thought. Instead, something in the lysogenic cell actively enforces its slumber.

This was a fact to file away—a piece of information that might serve some purpose one day. But for now, Jacob had no idea what the molecular sleeping pill might be, or how it might act. The ways of the phage were as mysterious as ever.

ANOTHER KIND OF CHOICE

ONE NEEDED A THICK SKIN TO SURVIVE THE ATTIC—THE EXCHANGES were passionate and sometimes personal. But the exercise had a purpose, for with every barrage of criticism came an idea for a new experiment, something that could confirm or refute a hypothesis. Each argument meant another possibility that had to be ruled out, a fresh analysis, a way forward. The speed with which bacteria and phage reproduce—doubling in less than an hour—lent itself nicely to a daily routine: debate at lunchtime, set up an experiment in the afternoon, review the results the next morning, and present the new findings to the group over lunch. A cycle of hypothesis, refutation, and refinement.

At the opposite end of the hall from Jacob and Lwoff, another microbiologist—a man 10 years Jacob's senior—was preoccupied with the eating habits of bacteria. Jacques Monod had completed his scientific training before the war, sparing him some of the battlefield ordeals and tortured wanderings Jacob had endured. Monod's initiation into the world of science was far more leisurely, the years between completing his bachelor's degree and starting his PhD spent sampling science, music, and sailing in nearly equal quantities.

As a graduate student, Monod had taken an interest in the response of bacteria to different diets, especially carbohydrate sugars. When the bacteria were supplied with only a single sugar, either glucose or lactose, they consumed it much as one might expect, dividing exponentially until the food supply was exhausted. But when Monod gave bacteria both glucose and lactose at the same time, an interesting pattern emerged: the bacteria grew, then paused, then grew again, a pattern he termed *diauxie* (from Greek *auxein*, to grow, and *di*, two).

Monod speculated, and later confirmed, that the different parts of

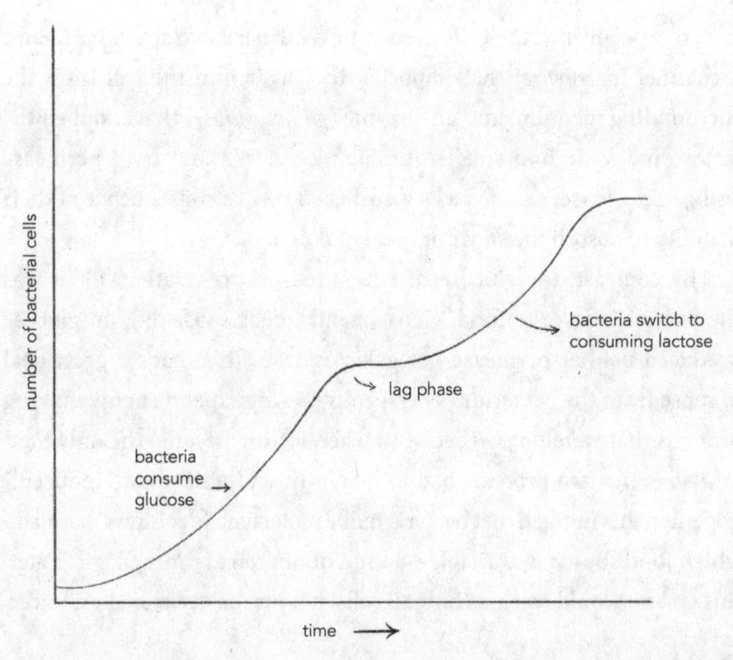

Given a mixed diet of two sugars–glucose and lactose–bacterial cells will first consume the glucose and then, following a change to their metabolic programs, switch to lactose. The growth pattern resulting from these dietary "decisions" is called diauxic growth.

the growth curve corresponded to the utilization of different sugars. Initially, the bacteria consumed the glucose, fueling the first phase of the curve. Then, once the glucose was gone, there was a resting phase, after which the bacteria started consuming lactose until this too was gone. But it was the resting period—the interlude separating the two growth phases—that Monod found especially interesting, as it indicated that a molecular switch was governing the transition from one sugar to another. Rather than devouring their meal in a gluttonous frenzy, bacteria seemed to eat one course at a time, *choosing* what to devour first and what to devour second.

Studying the problem in some detail, Monod found that the bacteria had to make some additional preparations before they could consume lactose. Specifically, the cells needed to synthesize two special proteins: a channel (*permease*) that imported the sugar into the cell from the surrounding medium, and an enzyme (*galactosidase*) that chopped the lactose molecule into smaller, usable pieces. Without both permease and galactosidase, cells fed a lactose diet starved, a consequence of their inability to absorb the sugar or break it down.

By contrast, the glucose diet required no preparation, for it was the bacteria's favorite food. Consequently, cells subsisting on glucose produced neither permease nor galactosidase. This made a great deal of sense from the bacterium's perspective—why expend energy making proteins that were not needed? But when lactose became the only food available, the two proteins had to be newly synthesized, or "induced." Together, this implied that bacteria make molecular "decisions" based on which food source is available—a kind of microbial dining logic. Later, this system would teach us how all cells, not just bacteria, make choices.

IN THE SUMMER OF 1958, JACOB'S THOUGHTS KEPT RETURNING TO THE phage. He had grown weary of preparing lectures about conjugation and the travel required to deliver them—the life of an academic. To be sure, the previous four years had been good to him. His work with Wollman had blossomed, and he had managed to build a nice reputation for himself. But something nagged at him. He had neglected the thing that led him to Lwoff's laboratory in the first place—the secrets of the phage—and his studies of bacterial conjugation had done little to advance that goal.

Perhaps that is not an accurate portrayal, however, for it neglects that one result he got during his early days working with Wollman—the conjugation experiment that showed that it was a sleeping pill, rather than the absence of an alarm clock, that kept the phage in its dormant state. The finding hinted at some undiscovered system of molecular regulation, one or more go-betweens interposed between the genetic script and its cellular output. At the time, Jacob had been too engrossed in his work with Wollman to take these observations further, but now he was ready for a new scientific challenge and, perhaps, a new collaborator.

By chance, Monod was also hitting a wall in his work. Joshua Lederberg, the American scientist who first discovered bacterial conjugation, had shipped several mutant strains of *E. coli*, and Monod was struggling to make sense of them. Each of Lederberg's mutants carried a defect in a single gene—a mutation that rendered it unable to digest one or more sugars. Monod was particularly interested in mutants that could no longer survive on a diet of lactose, the so-called "*lac* mutants." He believed that these strains might hold the key to understanding why bacteria engage in two-course meals when they could just as easily have consumed everything at once.

Lederberg's approach marked an evolution in the use of mutation to study biology. The earliest mutants—those obtained by Hugo de Vries and Thomas Hunt Morgan—had established many of the core principles of genetic inheritance. But by the 1940s, even as DNA was being revealed as the molecule of heredity, microbiologists had realized that mutants could also be used to study cellular and organismal function. If one examined carefully enough the consequences of a gene's absence, this would reveal something about that gene's normal role. It was the molecular equivalent of understanding how a house was electrically wired by unscrewing the fuses one at a time.

Monod began by characterizing Lederberg's *lac* mutant genes, later named *lacY*, *lacZ*, and *lacI*. It turned out that two of the mutants (*lacY* and *lacZ*) were incapable of digesting lactose at all, while the third mutant (*lacI*) digested lactose straightaway, without a lag in growth. The molecular basis for these behaviors was a complete mystery. Armed with an antibody that could measure the abundance of the galactosidase protein, Monod resolved to understand the functions of these three genes.

At first, he made good progress. Two of the mutant genes, *lacY* and *lacZ*, turned out to encode the two proteins whose function he had previously shown to be essential for lactose consumption: the *lacY* gene encoded the permease transporter, while the *lacZ* gene encoded the galactosidase enzyme. This made the interpretation of these two mutant strains relatively straightforward: neither mutant could survive on a lactose diet because they lacked the means to ingest, or digest, the sugar.

But this coherent picture of gene, protein, and function came unraveled as Monod strove to understand the workings of the third mutant—the *lacI* strain. This was the strain that had been able to subsist on lactose without needing the presence of the sugar to induce its metabolism. Using his antibody that could recognize galactosidase, Monod showed that the enzyme was always present in *lacI* mutant bacteria, even in the absence of lactose. This made little sense. All of Monod's previous studies had indicated that lactose could be metabolized only if it first induced the means of its uptake and breakdown. In *lacI* mutants, no induction was required.

Jacob, who had listened to Monod's daily progress reports over lunch, saw a similarity with his own work. It occurred to him that maybe the normal role of the *lacI* gene was to hinder the synthesis of galactosidase. In other words, perhaps *lacI* was acting as an inhibitor, much as his experiments with phage had hinted at an inhibitory sedative. Perhaps both scientists were dealing with inhibitors—in Monod's case, an inhibitor that prevented the cell from digesting lactose, and in Jacob's case, an inhibitor that prevented the prophage from awakening.

Jacob approached his older colleague with a proposition: As neither was equipped to study the question alone, why not work together? Doing so—merging Monod's collection of mutant bacteria with Jacob's conjugation techniques—might give them a fighting chance of solving the puzzle of these enigmatic molecular barriers.

JACOB WATCHED AS MONOD OPENED THE DOOR TO THE FREEZER. Inside were shelves stacked with frost-covered boxes, home to the mutant bacterial stocks Monod had labored to understand for more than a year. After a brief search, Monod found the one he was looking for. It held a dozen finger-sized tubes, each inscribed with a notation identifying its contents—a scribble that only Monod could decipher. These the microbiologist inspected until he found two that seemed to be the ones he was looking for. Satisfied that he had what he needed, Monod returned the box to the freezer and walked to the laboratory bench where Jacob was waiting.

Now, it was Jacob's turn. He produced a thin metal rod with a small wire loop attached to the end. He sterilized the loop by holding the tip over a small flame, removing any last traces of bacteria that might have remained from a previous experiment. Jacob then uncapped one of the frozen tubes that Monod had retrieved and plunged the hot loop into the icy bed of frozen material inside, thawing its surface. A tiny quantity of the melted liquid, less than a droplet, clung to the loop. This was more than enough to inoculate a fresh culture, for every microliter of fluid held millions of bacteria, still viable in their frozen state. Jacob stirred the loop into a broth-filled flask and then repeated the whole procedure on the second tube Monod had carried over. After a few hours in an incubator, the cultures would be turbid and ready for conjugation, the microbial mating ritual for which this entire exercise had been preparation.

It was 1959, and Jacob and Monod's partnership was thriving. The two Parisians had been joined by a third scientist—an American biochemist on sabbatical from the University of California at Berkeley named Art Pardee—and the trio made for a formidable brain trust. In addition, the researchers had made some technical advances. The most important of these was a chemical analog of lactose called ONPG, which turned from colorless to yellow in the presence of galactosidase. This made it much easier to measure the enzyme's activity, allowing the researchers to measure its accumulation during the process of induction.

Still, each experiment involved a week of planning, and this forced the scientists to decide which permutation should take priority. They agreed to begin with a control, a cross between two strains: (1) a normal, or wild-type, male in which galactosidase was produced only in the presence of lactose, and (2) a female lacking the genes for both galactosidase (*lacZ*) and the mysterious inducer (*lacI*). Or, as notated in the arcane nomenclature of the Attic:

$$♂ lacZ^{+} lacI^{+} \times ♀ lacZ^{-} lacI^{-}$$

Jacob, Monod, and Pardee worked in silence, collecting samples from conjugal pairings at precise 10-minute intervals, which they then brought downstairs to measure galactosidase levels with ONPG. Within two hours, the results were in. Galactosidase appeared rapidly following the mixing of the two strains, reaching its maximal rate of synthesis within minutes of the onset of conjugation. But after half an hour, galactosidase production ceased, after which it could be reactivated only upon the addition of the inducer, lactose.

From this seemingly innocuous result, Jacob knew, immediately, that things would never be the same.

THE DISCOVERY OF mRNA

IT IS RARE FOR AN EXPERIMENT TO BE GIVEN A NAME. BUT SUCH WAS the importance of the research carried out by Pardee, Jacob, and Monod that it became an exception, living on as the "PaJaMo experiment" in honor of its three architects. While the design of the experiment was simple, its interpretation is a taxing mental exercise. (For Jacob and his colleagues, immersed as they were in the details of mutants and crosses and inducers, the implications were easier to interpret.)

The PaJaMo experiment had several takeaways. First, it showed directly that genes can be regulated. The prevailing view when the scientists began their work had been that lactose directly interacted with galactosidase and permease. A plausible model, for example, might have been that galactosidase existed in the cell all along, but in an inactive form, whereupon a chemical interaction with lactose caused it to become active. But like Jacob's earlier experiments involving zygotic induction of the phage, the PaJaMo experiment showed that induction was happening at the level of the galactosidase *gene*, for it was the gene, not the protein, that had been transferred via conjugation. The gene itself had been the source of the rapid induction.

The second conclusion had to do with the way the gene was being regulated—a process that came to be known as "repression." Monod had known for a while that *lacI* was responsible for inhibiting the production of galactosidase and permease, but he had been unable to figure out how it did so. Now, there was some clarity. Conjugation had caused a wild-type genome, containing both *lacZ* and *lacI*, to enter a cell that was missing them both. Upon transfer, galactosidase (the product of *lacZ*) could be rapidly produced because the inhibitor (the product of *lacI*) was absent. But with time, the inhibitor (whose gene was also introduced via conjugation) was also made, accounting for the cessation of galactosidase synthesis a half hour later. In other words, the *lacI* gene seemed to be exerting its inhibitory effects through direct interaction with the *lacZ* gene rather than its protein product. How this occurred remained mysterious.

Finally, the third takeaway pertained to an inferred missing link between genes and proteins. Years earlier, George Beadle and Edward Tatum had shown that genes exist in a one-to-one relationship with proteins, such that each protein is encoded by a single gene and each gene encodes a single protein. But this landmark discovery left unsettled how information got converted from genetic blueprint (written in the language of nucleotides) to protein product (written in the language of amino acids). A new paradigm was needed to explain this transfer of meaning, and PaJaMo implicated a go-between, a hypothetical messenger serving as intermediary between the two.

IF THE PRECEDING DISCUSSION WAS HARD TO ABSORB, DON'T FRET; IT also took a while for Jacob to process the findings. There was no doubt in his mind that the results were significant, but assembling the pieces into a coherent picture of gene regulation—the molecular connection between a gene and its protein product—presented a huge challenge. Jacob filtered through numerous possibilities in his mind, considering how various models might accommodate the results. And each time, he returned to the middleman hypothesis.

In the fall of 1959, an opportunity to test-drive these ideas presented itself—a scientific meeting in Copenhagen whose attendees would include the leading biochemists and geneticists of the day. Jacob's ideas were still rough-hewn, but he felt bold enough to stick his neck out. He reasoned that the best way to determine whether his go-between model had any merit, or was fatally flawed, would be to follow the same practice he had observed all along in the Attic: to present his ideas to the best minds and let them have at it.

In front of a packed room, Jacob proposed the existence of a short-lived and unstable "messenger"—a mystery molecule he called X—to explain the rapid rise and gradual decline in galactosidase levels from the PaJaMo experiment. It was a hypothesis born of intuition, based largely on circumstantial evidence. So, naturally, Jacob expected the crowd to tear the idea apart with questions and criticisms.

Instead, there was silence—no agreement, but no outcry either.

In the audience, Francis Crick and Sydney Brenner listened intently. Crick—the intellectual powerhouse who had helped define the structure of DNA six years earlier (the famous "Watson-Crick" double helix)—was already a celebrated biologist, while Brenner, whom we will get to know better in the next chapter, would have to wait a few years for his fame to arrive. But at that moment, both men were focused on the same question that had stymied Jacob: How do instructions in DNA result in the synthesis of protein? Crick asserted that genetic information could flow in only one direction, from DNA to protein and not the reverse. But this proposition, which he called "the central dogma," failed to explain how the information moved. Despite its holes, and a reliance on a phantom X intermediate, Crick and Brenner had to admit that Jacob's proposal had merit. They wanted to hear more.

The following spring, Brenner hosted a gathering at Cambridge University, where he had just taken a senior research position. Jacob was invited, as were Crick and other prominent biologists. The gathering took place in Brenner's rooms at King's College, and the atmosphere was casual, allowing the participants to speak their minds openly and without judgment. All relevant facts, experiments done in various labs throughout the world, were laid out for discussion, like pieces of evidence in a murder case. The question at hand: What, if anything, lies between a gene and its protein product?

A model began to emerge within the group, and amazingly (to Jacob), molecule X was its centerpiece. During the dialogue, it became clear that a unique property of X was that it must be produced rapidly and destroyed just as quickly, a feature that distinguished it from most other substances in the cell. Brenner and Crick looked at each other, realizing that a substance possessing these features had just recently been described—a molecule that appeared as soon as bacteria were infected with a bacteriophage similar to, but distinct from, phage λ. The molecule in question was a special type of ribonucleic acid (RNA), a chemical relative of DNA that had been ignored because it constituted such a tiny fraction of the total RNA in a cell. But the molecule behaved just as the X should, and Brenner and Jacob gave it the working name "messenger RNA."

As the discussion proceeded, the two scientists realized that they would both be spending the upcoming summer at Caltech—Jacob at the invitation of Max Delbrück, and Brenner at the invitation of new faculty member Matt Meselson. As the day came to a close, and other scientists retired to Brenner's living room for drinks and music, Jacob and Brenner stayed behind, plotting out the experiments that would expose messenger RNA as the middleman between DNA and protein.

In the summer of 1960, Jacob and Brenner made their way to Southern California, where Meselson had been preparing for their arrival. After hearing of Jacob's interesting hypotheses, Meselson had worked to refine some of his techniques—methods that could confirm or refute the existence of X and further characterize its nature. The days were filled with ups and downs, periods of disappointment followed by bliss. But by the time September rolled around, the scientists had all the evidence they needed. Messenger RNA, or mRNA, as they began calling it, had earned a firm place in Crick's central dogma, nestled between a gene and its corresponding protein.

THE REPRESSOR

AS IMPORTANT AS THESE FINDINGS WERE, THEY EXPLAINED ONLY part of the PaJaMo result—the fact that protein induction must occur through an mRNA intermediate. PaJaMo's other feature—that regulation of the gene occurred through an inhibitor—remained as puzzling as ever. Jacob's mind went back and forth between phage and *E. coli*, looking for an explanation of how genes and their corresponding proteins might be regulated in these two seemingly disparate systems.

Conventional wisdom held that inducers acted directly on proteins to stimulate their synthesis or activity. But Jacob knew that there was something wrong with this explanation. The phage and bacteria were poised to produce proteins all along—like a car on a hill held in place by a molecular brake, a *repressor.* In bacteria, *lacI* acted as this molecular brake, blocking the production of galactosidase and permease. But when lactose was present, the brake was released. In λ, an inhibitor whose identity was still unknown blocked the production of viral proteins, keeping the phage in its lysogenic snooze until the inducer of the prophage—UV light—managed to release the brake on the phage's lytic cycle. And if the role of the inducer was simply to inactivate the repressor, then induction was not really "induction" after all—it was a release from repression. Jacob, and everyone else, had been looking at the problem upside down.

All this was conceptually helpful, but it still didn't explain how a repressor worked at a molecular level—how it interfered, in chemical terms, with the production of a protein from its corresponding gene.

Struggling to answer this question, Jacob found himself once again at a conceptual roadblock.

WE TALKED EARLIER ABOUT HOW DIFFICULT IT IS TO "UNKNOW" A fact—to forget, even temporarily, that the body is made up of cells or that a sand dune is made up of grains. The same is true when it comes to the molecular mechanisms of gene regulation. For the challenge facing Jacob was to understand how a gene gets turned ON or OFF without access to a critical piece of information: that the DNA molecule—the physical embodiment of a gene—is a participant in its own regulation. It is yet another fact that I cannot force myself to forget, yet in 1960 this thought was inconceivable. Virtually every biologist and chemist considered DNA to be untouchable—"a sacrosanct object that could not actually be manipulated without attacking life itself." Instead, it was assumed that proteins were synthesized from a DNA template one molecule at a time, a laborious process akin to a lithographer making limited-edition prints from a single master.

Jacob knew this had to be false. The speed with which the galactosidase enzyme was induced in the PaJaMo experiment was incompatible with such a slow and serial process. Bacteria could make too much protein in too short a time after induction, a finding more consistent with a model in which copies were being made from copies. Adding to the puzzle was the newly discovered go-between—mRNA—that lived between gene and protein, which also deserved a place in the gene-regulation framework.

But what ultimately pushed Jacob in the right direction was a feature the phage and bacterial systems shared, which had so far received little attention—the fact that multiple proteins were produced simultaneously. In *E. coli*, induction (the addition of lactose) caused permease and galactosidase to be produced with nearly identical timing. In λ, the synchronicity of induction was even more dramatic; dozens of proteins, not just two or three, came to life when the sleeping phage was induced to wake up. Again, this didn't make sense if proteins were synthesized one at a time from their corresponding genes.

It was in the most unexpected of circumstances, while watching a movie with his wife Lise in a theater, that the explanation came to Jacob in a flash of insight. It was a true Aha! moment—"the astonishment of the obvious," as he would later recall. There was only one place where the repressors could be acting in such a coordinated fashion: the DNA itself! The conceptual barrier that had held him back at last had a crack in it.

"I think I've just thought up something important," he told Lise as they left the theater.

Now, Jacob had to rebuild the entire system from this new vantage point. The ideas came in a flood, but he could finally think about the repressor in the correct way. Plus, there was now a place for mRNA to fit in: if the repressor could interact with DNA, then it could certainly inhibit the production of its corresponding message. A model began to take shape: if a gene was OFF, it was because a repressor blocked the synthesis of its mRNA. If a gene was ON, it was because the repressor was absent, allowing its mRNA to be made. A simple switch to regulate the *expression* of a gene. The model needed refinement, and Monod was the perfect test case. He was sure the older scientist would see the theory as heresy, a concept that contradicted DNA's (presumed) untouchable nature. Jacob knew that if he could convince Monod, then he could also convince the world.

As expected, his partner was resistant. But Jacob persisted, returning to Monod once, twice, three times, just as he had with Lwoff. Finally, his colleague's brow started to furrow. Monod was becoming curious. In response to each of his colleague's objections to the repressor model, Jacob rescued it with another piece of logic. Bit by bit, Monod began to grasp the essence of the theory and, more importantly, its elegance. The debate was no longer between Jacob and Monod; now it was Monod versus Monod, as the scientist attacked and defended both sides of the question, balancing arguments in favor of the model with arguments against. It was only a matter of time before Monod would come to his side. Jacob knew he had won.

CHEMICAL CIRCUITS

IN THE MONTHS THAT FOLLOWED, JACOB AND MONOD REFINED THE model with a series of confirmatory experiments, building a framework for understanding the principles of gene regulation. The essence of their model still holds true today, and even though the details were first worked out in bacteria and phage, the fundamentals extend across all life on earth.

The first step in gene regulation is *transcription*, the process by which DNA gets converted to its mRNA messenger. Transcription exploits the elegant economy of nucleic acid "base pairing," the chemistry that holds the two strands of the DNA double helix together by virtue of the mirror-image chemical affinities that certain bases have for their complementary partners: cytosine (C) is drawn to guanine (G), and adenine (A) is drawn to thymine (T). For example, if one strand of DNA contains the sequence GAATTC, its counterpart on the other strand of the double helix must be CTTAAG. During transcription, the two DNA strands separate, allowing an enzyme known as RNA polymerase to squeeze in between and "read" one of the DNA strands. The result is an mRNA molecule that is a near mirror image of its DNA template, akin to a photographic print made from a film negative.

The essence of Jacob and Monod's model is that repressors regulate transcription, the synthesis of mRNA from a gene, by binding to specific DNA sequences. In the OFF state, the repressor acts like a nightclub bouncer, ejecting RNA polymerase should it get close to the gene it is guarding. In the ON state, the doors are open, and RNA polymerase is free to act, resulting in an abundance of mRNA messages.

For a more concrete example, we can return to the diet of *E. coli*, where lactose induces cells to make the proteins allowing the sugar to be ingested (permease) and digested (galactosidase). In the absence of lactose, a repressor protein—the product of the *lacI* gene—binds to the DNA sequences encoding permease and galactosidase—*lacY* and *lacZ*—and blocks RNA polymerase. Consequently, no mRNA is produced. But

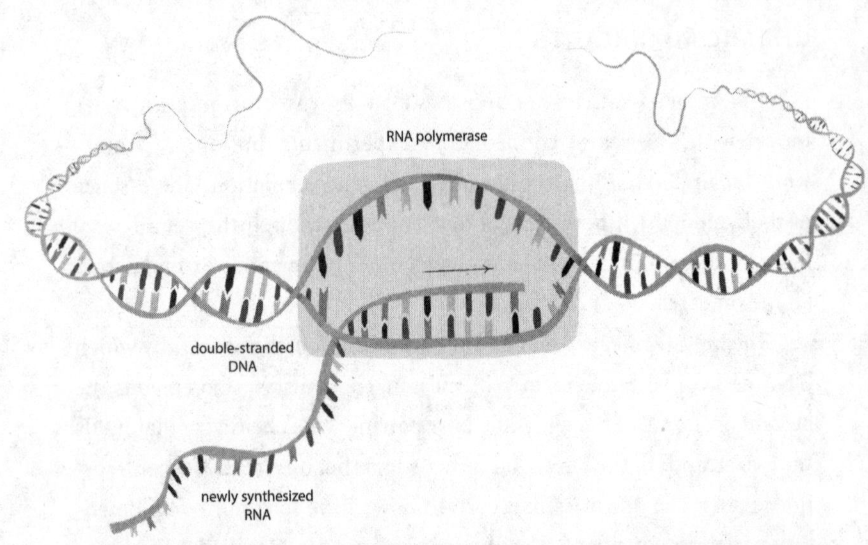

"Transcription" refers to the production of a message (mRNA) from a DNA template. Once the two strands of the DNA double helix have separated, the enzyme RNA polymerase synthesizes a mirror-image copy from one of the two strands. Sequences embedded in the DNA provide instructions to RNA polymerase so that transcription begins and ends at the appropriate positions. Once it has been synthesized, the mRNA molecule is transported from the nucleus to the cytoplasm, where it can be translated into protein.

when lactose is present, even in small quantities, the sugar associates with the repressor, pulling it away from the DNA. Now, RNA polymerase can reach the *lacY* and *lacZ* genes, resulting in many mRNA copies. (A similar system operates in phage, where a virally encoded protein binds to DNA sequences in the phage genome, preventing the synthesis of mRNAs necessary for viral replication. Exposure to UV light—Lwoff's "induction of the prophage"—destroys the repressor, allowing these genes and their lethal protein products to be expressed.)

Jacob and Monod's model opened a Pandora's box of new questions: If repressors regulate transcription by turning genes OFF, then might there be other proteins—*activators*—that can turn genes ON? (*There are.*) Do certain DNA sequences serve solely as docking sites for repressors and activators, having nothing to do with a gene's protein product?

(*They do.*) Is there a counterpart of this microbial system of gene regulation in animals? (*There is.*) Does this model explain what allows cells to take on differentiated identities in the face of a single, invariant script? (*It does, in part.*)

It turns out that all organisms—from plants to animals—devote tremendous energy to the regulation of transcription. In addition to the many *transcriptional repressors* encoded by the genome, cells contain a comparable number of *transcriptional activators*, whose role is to turn a gene ON by augmenting the activity of RNA polymerase. The protein repressors and activators that regulate transcription are referred to collectively as *transcription factors*, and their sole purpose is to give voice to a gene or render it silent. The human genome contains as many as 1,500 transcription factors, comprising 5 to 10 percent of the entire genome. This is a surprisingly high number, given that the chief purpose of these proteins is to regulate the production of other proteins.

If we take a step back, this investment in regulation is essential for life. A static genome, with fixed instructions, would leave a cell incapable of responding to its environment or communicating with other cells. The plasticity we saw in earlier chapters—the embryo's ability to cope with cellular loss or the events that give rise to the cellular society—would be impossible without a flexible genome. Regulation of transcription is the most prevalent and most ancient way for a cell to adapt to changing circumstances. Consequently, almost every biological process imaginable—from growth to repair and from sensation to memory—is influenced by the rate at which different DNA sequences are converted into mRNA.

TRANSCRIPTION IS JUST ONE WAY FOR CELLS TO REGULATE INFORMAtion in the genome. The next step is *translation*, the conversion of the string of nucleotide bases in an mRNA to the string of amino acids in a protein. The conversion from mRNA to protein is even more complicated than the conversion from DNA to mRNA, for while the nucleotide alphabet of DNA and RNA has only 4 letters, the amino acid alphabet of proteins has 20. Consequently, base pairing—which enables the precise synthesis of mRNA from a mirror-image DNA template—plays a different role in translation. Indeed, the term "translation" is particularly apt, since the formation of an amino acid chain from an mRNA molecule requires, in molecular terms, a linguistic overhaul.

In the mid-1950s, physicist George Gamow proposed a model for this molecular leap from the nucleotide sequence of a gene to the amino acid sequence of its corresponding protein. Gamow reasoned that a system, or code, that interpreted nucleotide bases clustered in groups of three (for example, AAA, ATA, ATT, and so on) could be combined in 64 possible ways ($4 \times 4 \times 4$), more than enough to encode the 20 different amino acids that constitute the building blocks of proteins. This was seemingly the best fit, for a code based on two nucleotides would fall short of the necessary 20 possible permutations ($4 \times 4 = 16$), while a code based on four or more nucleotides would result in a seemingly wasteful number of permutations ($4 \times 4 \times 4 \times 4 = 256$).

Indeed, three is the magic number when it comes to translation. After a gene has been "transcribed" into its corresponding mRNA molecule, another enzymatic complex, the "ribosome," begins to scan the message, looking for a particular three-letter word that tells it to start making protein. Each triplet is called a *codon*, its three-letter combination prescribing which amino acid should be added to the growing protein chain. It took only a few years following Jacob and Monod's discoveries for a group of scientists including Marshall Nirenberg, Heinrich Matthaei, Har Gobind Khorana, and Philip Leder to link the meaning of each three-letter codeword in DNA to its corresponding amino acid. The result was a living Rosetta stone known as the "genetic code," a constant of nature as central to biology as Newton's laws of gravity are to physics.

The elucidation of the genetic code marked the birth of molecular biology, for the cipher made it possible to infer the amino acid sequence of any protein from the DNA sequence of its corresponding gene. Bio-

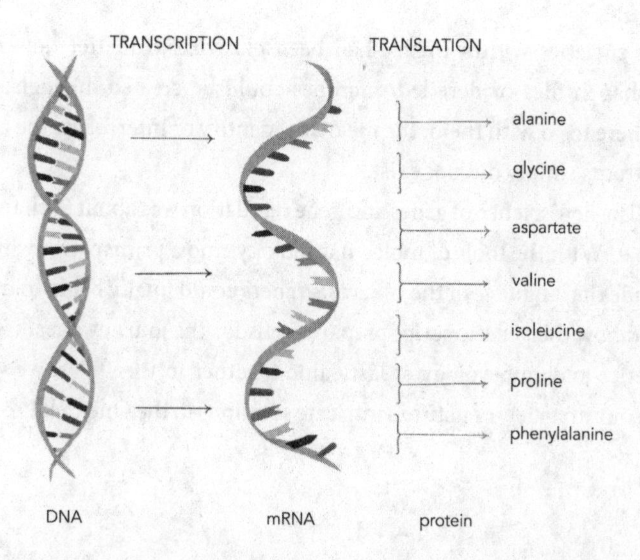

Genes encrypt protein-coding information in three-letter codewords known as codons. Once a gene has been transcribed into mRNA, another molecular machine–the ribosome–converts each three-letter word into one of the 20 amino acids that make up proteins. Each amino acid is added sequentially, forming a string whose sequence is unique for each protein. Three codons (TGA, TAA, and TAG) do not code for any amino acid but rather tell the expanding chain of amino acids when to stop growing. Such "termination codons" serve as periods at the end of protein sentences formed by amino acids.

chemists developed better and faster techniques for sequencing and manipulating DNA, providing an unprecedented window into the inner workings of the cell. (A gene's DNA sequence can, at present, be determined more rapidly than a protein's amino acid sequence.) And everywhere they looked, scientists found that even subtle molecular differences—changing just a single base and hence a single amino acid—could have profound effects on proteins, cells, tissues, and organisms.

For generations, genetic diversity had been the biologist's greatest tool and teacher. It accounted for the range of species that launched Darwin's quest and the variation in plant size and color from which Mendel formulated the laws of heredity. But the grounds for that variation—the triggers that spawned the diversification of species or the untold cellular societies within them—had been elusive. No matter how many misshapen flies or derailed microbes could be created, biologists had nowhere to go with them; the means to identify or interrogate the causal apparatus simply did not exist.

The new science of genes and gene regulation was about to change all of that. With the tools of molecular biology, and a primer with which to decode the language of the cell, researchers could finally begin to understand how the genetic script prepares cells for the journey ahead. When genetics and embryology at last came together in the 1970s, we would gain our first view of nature's intricate solutions to the One Cell Problem.

Chapter 5

GENES AND DEVELOPMENT

You have made your way from worm to man,
and much of you is still worm.

— FRIEDRICH NIETZSCHE, *THUS SPOKE ZARATHUSTRA*

There are two sides to the developmental coin: the *What* and the *How*. The What is descriptive: it paints a picture of whatever cells do as they contribute to their cellular societies. The How, by contrast, is concerned with root causes: the chemical and physical programs by which the marvel of embryogenesis plays out. The What is about the things we see, while the How is about mechanism—the shadowy ghost-in-the-machine that takes inert elements and breathes life into them. It was the How that Aristotle had in mind when invoking "entelechy," the almost-mystical force that converts the potential into the actual.

Before the late nineteenth century, naturalists and philosophers focused almost exclusively on the What, arranging the natural world into neat categories. The early pioneers of experimental embryology—Roux, Driesch, and Spemann—only nibbled at the How, uncovering bedrock principles of development that clarified the path from embryo to maturity. But it was only with the next generation of scientists—researchers who were as interested in the chemistry of life as they were in its biological manifestations—that the How of development came into the spotlight. It didn't matter that microbes, which do not undergo embryogenesis, had served as the principal source of their discoveries, for they knew that animals would prove to be intimately linked to their unicellular forebears.

"What is true for *E. coli* is true for the elephant!" Monod had declared.

The elucidation of the genetic code in the 1960s had given molecular biologists access to the language of the cell. But at first, there was hardly anything to read. The experiments that spawned biology's central dogma—that information flows from DNA to mRNA to protein—had relied on laboratory-synthesized nucleic acids. But to make sense of development, biologists needed the real thing: the naturally occurring DNA sequences that propel cells to their roles and positions. They would have to wait a little. The revolution in "recombinant" DNA technology—an experimental toolbox that would allow scientists to isolate, sequence, mass-produce, and transport whatever pieces of DNA or mRNA they wished—was still more than a decade off.

The term "genetics" can mean different things to different people. In the broadest sense, it refers to anything related to heredity or DNA. When a son has his mother's eyes, we may attribute that trait to "genetics." The same goes for when a cancer "runs" in a family. In addition to similarities, the word can also be used (or misused) to highlight differences. Traits such as race, body size, sexual orientation, personality, and criminal tendencies have all been incorrectly or imprecisely assigned a "genetic" foundation, providing a convenient way to isolate or degrade whole groups of people. Population geneticists concern themselves with the transmission of gene variants, or alleles, within a species, while molecular geneticists focus on the genetic material itself.

For our present purposes, "genetics" has a completely different meaning: it is the use of inherited variants—mutants—to study biology. In other words, genetics is an *approach*. The method can lead to unexpected biological insights, and its power and beauty come from the fact that it requires little a priori knowledge. The logic behind the genetic approach is simple: observe the anomaly to understand the normal. If a mutation in a gene (an altered genotype) causes a new or different trait (an altered phenotype), then the gene in question must be tied, in some way, to that trait. This, in turn, makes it possible to play biology in reverse—using mutant phenotypes as a lure with which to catch and unravel the molecular underpinnings of an opaque cellular process.

Consider Thomas Hunt Morgan's groundbreaking discovery of the *white* mutation. Morgan took years to find his special fly, and it provided formal proof that genes have a physical presence on chromosomes. But Morgan's discovery also had a second, less appreciated, meaning. Because the *white* mutation resulted in colorless eyes, it implied that the *white* gene, in its unmutated or wild-type form, gives the fly eye its typical reddish hue. (Morgan lacked the tools to study *Drosophila* eye color in detail, but years later other researchers discovered that the *white* gene encodes a protein that carries red pigments into the developing fly's eye, explaining why its absence causes ocular albinism.) In short, genetics provides an answer, and it is up to the geneticist to understand what that answer means.

The genetic approach can lead anywhere, as its goal is not to test hypotheses but rather to generate new ones. Consequently, the outcome of a genetic expedition—an effort to find mutants with defined abnormalities—cannot be predicted, as it may lead to an orderly model or a chaotic jumble. The genetic approach is not for the fainthearted. But for those scientists with sufficient drive and vision, like the ones we are about to meet, genetics can pull away the curtain of mechanism, revealing the fundamental secrets of embryonic development and the One Cell Problem. It is an entry point into the How.

IN 1978, AS THE REVOLUTION IN RECOMBINANT DNA TECHNOLOGY WAS well underway, Eric Wieschaus and Christiane Nüsslein-Volhard were setting up their *Drosophila* laboratory at the newly created European Molecular Biology Laboratory in Heidelberg, Germany. Wieschaus was an American who had moved to Heidelberg at the age of 41 following a

PhD at Yale and postdoctoral work in Basel and Zurich. Raised in Birmingham, Alabama, the young Wieschaus had imagined his adult self as an artist. But a summer program at the University of Kansas during high school convinced him that science and research could be an alternative outlet for his creative impulses—different from the visual arts but drawing on the same desire to look at the world in new ways. Nüsslein-Volhard, five years older than Wieschaus, grew up in Frankfurt, the provisional capital of postwar West Germany. She developed a love of nature early in life, and by the age of 12 she had declared that she would become a biologist one day. Like John Gurdon, whose ambles through the fields of Surrey had filled him with a similar childhood love of nature, she was easily distracted at school ("decidedly lazy," is how one of her teachers put it). And also, again like Gurdon, it was in the laboratory, not the classroom, that she found her passions.

The two researchers met in Basel while studying with Walter Gehring, a renowned *Drosophila* biologist. Gehring suspected that once the ability to rapidly and affordably sequence genes became commonplace, it would transform the study of development. This was a shift from the approach of preceding fly geneticists, who had been content with identifying mutant insects and categorizing their abnormalities without much concern over the underlying molecular principles. DNA sequencing, Gehring predicted, would one day bridge the gap between the What and the How.

Wieschaus and Nüsslein-Volhard hit it off immediately. They shared Gehring's vision for exploiting genetics to understand development, and their backgrounds provided the complementary skills that would be needed to fulfill that vision. While Wieschaus was an experienced fruit fly biologist, he lacked a molecular background. Nüsslein-Volhard, by contrast, was an experienced molecular biologist—her PhD had focused on the RNA polymerase enzyme, the engine of transcription whose existence Jacob and Monod had anticipated. But she had grown weary of the molecules, detached as they were from living creatures. In Gehring's lab, with its focus on mutant genes, she found that she could apply her molecular expertise to the frontiers of embryonic development. When an opportunity presented itself for the two scientists to launch a joint lab in Heidelberg, they jumped at it.

The pair were particularly interested in *patterning*, the process by which cells find their way to the correct positions in the embryo. For half a century, scientists had identified many genes responsible for form—genes that when mutated caused misshapen wings, legs, or torsos. But Wieschaus and Nüsslein-Volhard sought a more comprehensive view of development—a molecular blueprint guiding the construction of a body. In the cramped hallways of Gehring's laboratory, and over long dinner conversations, they plotted their strategy for accomplishing this in their own laboratory. Naturally, their scheme involved the genetic approach, and they imagined that with it they might be able to identify *all* the genes governing the structure of the fly, not just a small subset.

This was a risky undertaking. The scientists were both in their 40s and had arrived at the make-or-break period of establishing a laboratory. It was a time that most would consider the most stressful point of an academic career, when every scientist, no longer under the experienced eye of a graduate or postdoctoral mentor, must prove their worth and independence. If they were unable to obtain momentum during this critical time—publishing papers and securing grants, the sources of oxygen for a research career—they would have little hope of a future. No one would have blamed the two assistant professors for playing it safe, working on something that would extend their postdoctoral studies in a more incremental fashion. But they had grander ideas in mind and would either go big or go home.

HEAD, SHOULDERS, KNEES, AND TOES

DROSOPHILA MELANOGASTER IS THE IDEAL SUBJECT FOR STUDYING animal development. The insect is complex enough to reveal features of embryogenesis that all animals share, while its simplicity facilitates detailed investigation. At a molecular level, flies and mammals share many similarities—evolutionarily conserved programs that guide the formation of cellular societies. Of course, there are also profound differences. One difference appears shortly after the egg is fertilized, while the embryo is still at its earliest stages. Unlike frogs or mammals, more evolutionarily "advanced" animals whose early forms arise through successive rounds of cell division ("cleavage divisions"), fly embryos do not exhibit an increase in cell number in the hours following fertilization; instead, only their nuclei reproduce. Because no membranous walls materialize to surround these multiplying nuclei, the early stages of fly development transpire entirely within the confines of a single cell. The result is a *syncytium*—a throng of nuclei crammed into the *cytoplasmic* space previously occupied by the egg. It is only once the embryo has amassed some 6,000 nuclei—the aftermath of precisely 13 nuclear divisions—that boundaries form between them, demarcating them as individual cells. Prior to this stage of cellular segregation, the embryo is known as the syncytial *blastoderm*.

Although lacking in any apparent structure, the blastoderm has an underlying organization to it. Each region in this seemingly amorphous bag of nuclei possesses "spatial information"—a sense of its position in space and its future role in the adult fly. Remarkably, this spatial awareness precedes the formation of cell boundaries. In other words, each nucleus's future is predictable even before the cell that will arise from it comes into existence. Nuclei at one end of the blastoderm will form the fly's crown, nuclei at the other end will form its genitalia, and nuclei in the middle will form its thorax and abdomen. The insect's full shape—its dipterous head, shoulders, knees, and toes, as it were—is foreshadowed in the nuclei composing the blastoderm.

At first blush, the existence of such a "pre-pattern" may seem a throwback to preformationism and Weismann's mosaic model, the outdated idea that an animal's form is already in place from the moment of conception. Scores of biologists, starting with Hans Driesch, had shown that development is regulated, not deterministic, which had caused the mosaic model to be largely abandoned as a theory by the late nineteenth century. Thus, the notion that the fate of a nucleus can be predicted based on where it lives within the *Drosophila* blastoderm would appear, at face value, to violate this principle.

To see our way out of this paradox, we must first understand the meaning of the word "fate" in the context of development. The *Oxford English Dictionary* defines fate as "the principle, power, or agency by which, according to certain philosophical and popular systems of belief, all events, or some events in particular, are *unalterably predetermined* from eternity." So, when Shakespeare's Julius Caesar asks, "What can be avoided whose end is purposed by the gods?" there is no confusing the answer. Fate, in literature as in everyday life, is taken to mean an unalterable preordination, the soothsayer's edict carved in stone. Fate describes the inevitable.

But in the embryo, fate has a different meaning. It is a default, the outcome that will happen if a tissue or a cell is left to its own devices. However, there is nothing inevitable about fate during development, for if circumstances should change—if a cell finds itself in a new environment—all bets are off. Embryonic fate therefore describes an expected trajectory, but one that is still malleable. It is, in essence, the same behavior we saw operating in Driesch's sea urchins and Spemann's newts. Whatever future awaits a nucleus in the syncytial blastoderm, that future remains conditional.

A bit later in development, once membranes have formed around each nucleus, the developing fly is a larva. (Think of insect larvae as juveniles or caterpillars that are beginning to prepare for metamorphosis, the conversion from embryo to adult.) Unlike the amorphous blastoderm, the larva is highly ordered, consisting of a head and 11 segments. Each segment has its own identity—the 3 segments closest to the head will form the fly's thorax and the 8 segments farthest from the head will form its abdomen. The cells that make up these structures lose their plasticity as they mature, and by the time the fly has reached the larval stage, most cells have become committed to a particular job in the cellular society.

Cells, like people, become set in their ways as they age.

THE HEIDELBERG SCREEN

IN THE NEARLY 70 YEARS AFTER MORGAN'S DISCOVERY OF *WHITE*, scientists had discovered dozens of *Drosophila* mutants. Each of these genetic abnormalities—causing eyes, wings, legs, and other body parts to have irregular shapes or colors—provided a new window into the secrets of the embryo. The most bizarre class of mutations, the so-called *homeotic mutations*, caused one part of the fly's body to be substituted for another, resulting in misshapen flies with an extra pair of wings or legs where their antennae should have been.

But when it came to understanding the genetic logic that gives an animal its shape, Morgan's way of finding mutants had serious limitations. For starters, it was tedious. Looking for mutants one at a time took an enormous amount of diligence and dedication. Moreover, the method could find only mutations that allowed the mutant insect to survive to adulthood. Any genes whose activity was critical for shaping the embryo at the larval stage, or before, would never survive long enough to be identified. Consequently, the genes that interested Wieschaus and Nüsslein-Volhard—genes that served as blueprints for the insect's basic structure—were unlikely to emerge from this standard method. If the Heidelberg scientists hoped to pinpoint the all-important genes regulating the embryo's pattern, they would have to go about it differently.

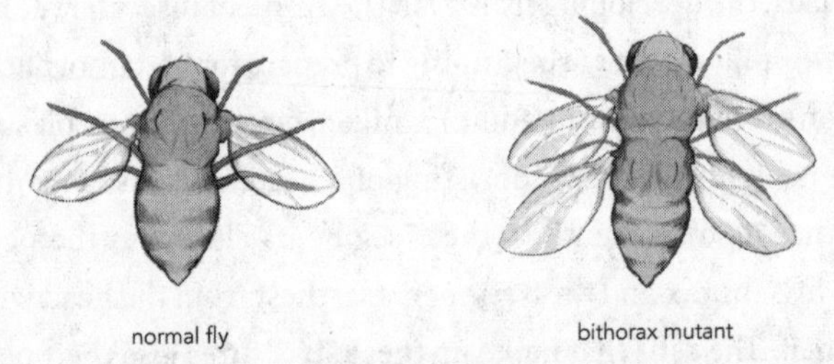

Homeotic mutations cause the substitution of one body part for another. In *bithorax* mutant flies, a segment of the thorax normally fated to become the haltere, an appendage used for balance, instead gives rise to a second pair of wings.

But this created a conundrum—how were they supposed to identify genes whose disruption would prevent the mutant flies of interest from ever being born?

The Heidelberg scientists contrived a clever solution. Rather than trying to find a way around this difficulty—the problem of *embryonic lethality*—they exploited it, devising an approach tailored to find mutants that didn't survive past the larval stage. The strategy was not perfect, as some genes resulting in embryonic lethality might have no role in shaping the embryo. (For example, mutating genes whose protein products have a more pervasive role in cellular sustenance would also result in death at early stages of development.) But the embryonic-lethal pool would also yield the kinds of genes Wieschaus and Nüsslein-Volhard wanted—genes that spell out how the fly achieves its form.

The first order of business was generating the mutants. Waiting for mutations to arise spontaneously, as Morgan had done, would be far too slow (recall the years Morgan spent waiting for his white-eyed fly to appear). Instead, the researchers used a trick developed by Ed Lewis, the *Drosophila* geneticist with whom they would later share the Nobel Prize. Lewis had found that exposing flies to ethyl methanesulfonate (EMS), a toxic chemical that attaches itself to DNA and converts G's to A's, was an efficient way to generate mutant sperm. Nüsslein-Volhard and Wieschaus would use this method, but instead of looking for adult phenotypes, as Lewis had done, they would use it to find mutations affecting fly larvae—especially those that couldn't develop any further.

As they prepared for their *genetic screen*, the two scientists could only guess at how many mutants they would need to filter through, but the estimates were staggering. The *Drosophila* genome contains more than 15,000 genes, and based on the laws of probability, they calculated that they would need to assess 40,000 independent mutant strains to be confident of mutating every gene at least once. Such a massive effort would require an array of new techniques. But at least the EMS approach would generate the necessary numbers of mutants, with thousands of mutant sperm created each week.

To manage a project on this scale, the researchers jerry-rigged a "high-throughput" system of tubes that allowed them to create, manipulate, and store dozens of mutant strains at a time. They applied other genetic tricks as well. The mutagenesis strategy, relying as it did on mutant sperm, yielded only *heterozygous* offspring (that is, flies carrying one mutant allele from the father and one normal allele from the mother). But as it was only the *homozygous* flies (that is, flies carrying two abnormal alleles) that were expected to exhibit a phenotype, the Heidelberg scientists had to interbreed the offspring of the first generation (F1) to generate second (F2) and third (F3) generations that would include such homozygous mutants. By preserving the previous generations during the analysis, keeping each stock separate, heterozygous mutants of interest could be recovered even if homozygotes died.

THE TRANSLUCENT CUTICLE SURROUNDING THE *DROSOPHILA* LARVA has a rather unassuming appearance, a chitinous, cigar-shaped shell whose accordion-like segments flex and extend like a medieval knight's suit of armor. As the larva develops, its newly formed tissues press into the cuticle, leaving an engraving-like impression of the forms below. So precise is this imprint that it captures the finest details of the juvenile, including its hairs and even its wrinkles. To the trained eyes of Wieschaus and Nüsslein-Volhard, the cuticle was a durable record of all embryonic mishaps.

The Heidelberg screen took nearly two years to set up but only a few months to execute—a brute-force approach requiring lots of tubes, lots of breeding, and lots of patience. The scientists did their best to automate as much as possible, but there was still one part of the screen that had to be done by hand: scanning the larvae for abnormalities. Now, the focus on embryonic lethality was critical, for it allowed them to home in on those mutant strains that never made it to adulthood and ignore the rest.

From that point on, it was just a matter of finding embryos that died in an "interesting" way.

The scientists sat across from each other at a two-headed microscope, a daily routine of rapidly scanning the mutant cuticles one by one in search of phenotypes. If the pattern was normal—the case for most mutants—the strain was discarded. If the pattern was abnormal, the strain was saved. Disagreements were rare, and when all was said and done, they had sifted through some 18,000 embryonic-lethal

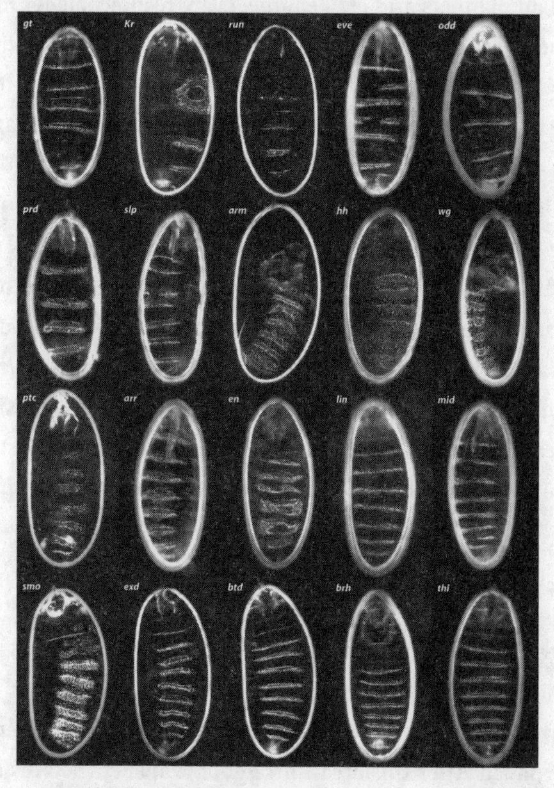

Mutations from the Heidelberg screen disrupted the cuticle-segmentation patterns of *Drosophila* larvae in various ways. For a sense of scale, each larva is three to four millimeters in length.

(REPRODUCED WITH PERMISSION FROM ANNUAL REVIEWS, INC.)

strains. Their reward: 120 genes that altered the shape of the fly cuticle in some meaningful way. In the tradition of the Fly Room, the Heidelberg mutants were christened with impressionistic names that reflected their unique deformed appearances: names like *hunchback* and *runt*—genes they hoped would prove to be "master regulators" of animal form.

In retrospect, it is remarkable that they found so few genes, as it is easy to imagine that a much larger portion of the fly genome could be dedicated to the insect's construction. As an example, think of the fly's body as a city, with its complex structures built from the ground up. There, urban planners and civil engineers are responsible for municipal development, overseeing such basic infrastructure as gas and electric, zoning, waste management, communications, mass transit, roadways, and food access. In embryos, the job of project management and infrastructure development falls to the patterning genes. But in contrast to cities, which may require the input of thousands of planners and engineers, the fly does the job with only 120 or so functionaries, all of whom complete their tasks in less than 24 hours. Rome may not have been built in a day, but this is how long it takes to build a *Drosophila* larva, its construction dependent on a small cadre of genetic supervisors. It could have easily turned out very differently. But nature, in its frugality, assigned the task of creating a body plan to a small leadership council, adjusting their individual responsibilities only incrementally across the sweep of time.

FROM FLIES TO WORMS

ACROSS THE ENGLISH CHANNEL, SYDNEY BRENNER WAS ALSO USING the genetic approach. We met Brenner earlier, when he helped François Jacob prove that mRNA serves as the intermediate between a gene and its protein product. After this breakthrough, Brenner turned his attention from molecules to embryos. But unlike the Heidelberg scientists, Brenner was more interested in behavior than shape.

Raised in South Africa, far removed from the exciting advances that were the domain of European and American biologists in the mid-twentieth century, Brenner nevertheless had an advantage: an uncanny ability to digest and remember everything he read. This autodidacticism, and his outsider status, allowed him to see connections that others missed. He attended medical school at South Africa's University of Witwatersrand but abandoned early on any thoughts of becoming a physician. The basic sciences, with their capacity to explain the fundamentals of human physiology, captured his imagination and passions. After graduation, he secured a research fellowship in England, where he would remain for the rest of his career. It was there (and during his summer at Caltech with Jacob) that he made his mark as a molecular biologist. But by the early 1970s, he had come to see the field as a rabbit hole of increasingly narrow questions. If one hoped to address the How of development, studying DNA and RNA in isolation could get you only so far, he thought. Brenner had helped bridge the gap between gene and protein product, and it was only fitting that he would next seek to connect gene with function.

Brenner's transition from biochemist to developmental biologist was thus driven by the same question that had propelled Wieschaus and Nüsslein-Volhard: How do genes control development? But his goal was more specific, and in some ways more audacious, than theirs, for he had set his sights on the most interesting and complex body structure of all: the nervous system. What was responsible for constructing and shaping that remarkable network of neurons, the instrument on which all of behavior is played? It must start with genes, he concluded, and with the right strategy, it should be possible to figure out what they were.

The choice of experimental organism was crucial. The microbes Brenner had studied as a molecular biologist would not do, as they lacked behaviors any more sophisticated than the eating and sleeping habits that Jacob and Monod had explored. But animals with complex brains and behaviors, like mammals or amphibians, were also inappropriate. Even flies—with their thousands of neurons—presented too daunting an experimental system for this kind of work. The ideal creature, he reasoned, should have a compact genome and a simple anatomy but still be advanced enough to register measurable behaviors. By 1970, Brenner had found his subject: an unassuming rod-shaped nematode roundworm with the unruly name of *Caenorhabditis elegans*.

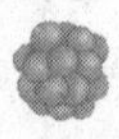

LITTLE WAS KNOWN ABOUT C. *ELEGANS* BEFORE BRENNER PLUCKED IT from obscurity. Transparent and just a millimeter long, the worm executes the behaviors associated with any animal—eating, moving, breathing, and reproducing. But unlike *Drosophila*, worms do not fly away if left unattended, and they cost almost nothing to keep, happily living out their three-week life span on agar-filled petri dishes. They boast a magnificent fecundity (a single animal can produce hundreds of offspring) and subsist on a diet of *E. coli*. As experimental animals go, they are on the low end of low maintenance. But what made *C. elegans* most attractive to Brenner was their behavior. The worms' movements were predictable: they slithered toward reward and away from danger, primitive activities that made them suitable for genetic investigation.

Not everyone at the world-renowned Laboratory of Molecular Biology in Cambridge—where Brenner had set up shop—shared his enthusiasm. Jokes about "Sydney's worms" ran rampant, and Brenner became the subject of respectful, if somewhat dubious, curiosity. The doubters had a point. Brenner could have spared himself an enormous amount of work by studying an "established" organism like the fly or the frog; instead, he was starting from scratch. But to Brenner, this was an opportunity. Genetic studies of behavior had to start simple, he reasoned, and the worm was as simple as it got.

Brenner's genetic approach was similar to that of the Heidelberg scientists, relying as theirs did on ethyl methanesulfonate to induce mutations, but with one major difference. Rather than looking for mutants with abnormal shapes, Brenner sought ones with abnormal *behaviors*, worms that no longer slid across the agar with the same elegance that had earned the species its name.

The mutants he found spun in endless circles, staggered instead of gliding, or simply sat in place, paralyzed, resulting in a total of 77 genes whose disruption led to a lack of coordination. Accordingly, he named his gene collection the *unc* mutants, for "uncoordinated," numbering them sequentially (that is, *unc-1*, *unc-2*, *unc-3*, and so on). The worm had yet to prove its worth as an experimental organism, a tool for discovery as suitable as the frog or the fly. But things were looking promising. The *unc* genes were required, in some way, for either the construction or function of the worm's nervous system, even if Brenner could not yet say how they worked. Determining what these and other developmentally important genes did—like the 120 shape-determining genes identified in the Heidelberg screen—would have to wait for technology to catch up.

THE FULL LINEAGE

BRENNER'S DESCRIPTION OF THE *UNC* MUTANTS, PUBLISHED IN 1974, made barely a ripple. Unable to characterize the genes responsible, he had to admit that his mutants were, for the moment, merely a curiosity—invertebrate movement disorders whose basis in physiology was unknown. But Brenner was not one to sit around, and there was still plenty to do. Some of the worm's most basic features were still a black box. For example, almost nothing was known about the worm's genome or its cellular anatomy. If *C. elegans* was to become as influential as *Drosophila*, as Brenner hoped it would, its genomic and cellular landscapes would need to be mapped.

Many believed Brenner was wasting his time. But the renowned Francis Crick was not among them. Crick had experienced Brenner's brilliance firsthand in the early days of mRNA, and he shared his younger colleague's interest in behavior and the nervous system. During a visit to the Salk Institute in La Jolla, California, Crick identified a fitting candidate to join Brenner's growing interest—an effervescent chemist named John Sulston. Over dinner with Crick, Sulston listened, skeptically at first, as the famous scientist described Brenner's scheme to use worms to understand behavior. The longer Sulston listened, the more his skepticism turned to fascination. Brenner's vision may have been farfetched, but Sulston was attracted rather than repelled by the idea.

"There's little point in doing what everybody else is doing," he reasoned.

Within days, it was decided—sight unseen, John Sulston would cast his lot with Sydney Brenner and the scarcely tested worm. A few months later, Sulston packed up his family, trading the pristine dunes of Southern California for the sun-hungry skies of southern England.

BEFORE SULSTON'S ARRIVAL, BRENNER HAD FOCUSED ALMOST EXCLUSIVELY on mutants with movement disorders—worms that staggered, stiffened, or otherwise displayed mobility problems. This early effort had left the worm's genome, the vault holding all the creature's instructions, untapped. The two agreed that Sulston, with his background in nucleic acid chemistry, would take on the worm's genome as his first project.

This proved to be rather straightforward. By chemical means—measuring with precision the average DNA content of a larva, then dividing by the number of cells per larva—Sulston calculated the size of the *C. elegans* genome to be approximately 80 million bases (80 megabases, or Mb), roughly 20 times that of the *E. coli* genome, half the size of the *Drosophila* genome, and 1/40th the size of the human genome.

As he got more comfortable with the worm system, something gnawed at Sulston, a simple fact that had seemed not to bother Brenner quite as much. It was cells, not genes, that made the worms move. While DNA sequences might bear the ultimate responsibility for shaping the embryo and its tissues, the task of pushing, pulling, stretching, and retracting—the act of repositioning—fell to the worm's cellular constituents. In contrast to the relative indifference of microbes to their neighbors, animal cells act collaboratively, dependent on vibrant conversations within their cellular societies. Such conversations may end with a shove or a tug, a straightening or bending, a reaching out, withdrawal, or retreat. And nowhere in the body are those cellular interactions more important than in the nervous system, where behavior is a dance of motor neurons, sensory neurons, and muscle cells. Studying movement without understanding the cells that trigger it, Sulston reasoned, might be as fruitless as studying a symphony without knowing the instruments that make up an orchestra.

Hoping to directly observe the cells mediating locomotion, Sulston shifted his attention to microscopy. His first hurdle was technical, a consequence of the very process of motion he wished to study. Microscopes are designed to visualize stationary items. But Sulston's subjects were living, growing, *moving* worms, whose natural tendency was to shimmy from side to side, flip over, wrap themselves in knots, or simply slither off the flat world of the microscope slide.

To solve this problem, Sulston embedded the worms in *E. coli*–laden agar. This kept them happy, as they could still eat and grow, but limited their mobility enough to keep them within the microscope's field of view. Another of his innovations was the addition of crosshairs to the microscope's eyepieces. This allowed him to follow the worms as they moved, just as World War II bombardiers had used crosshairs to keep their targets in view. Before he knew it, Sulston had transformed himself from chemist to worm neurologist, able to diagnose existing and novel movement disorders in his invertebrate subjects.

IF THE WORM WAS TO FULFILL ITS PROMISE, SULSTON BELIEVED THAT a detailed classification of all participating cells and their positions—a kind of cellular catalogue—would be a prerequisite. To make such a catalogue, he began mapping individual types of cells that could be identified based on the proteins they expressed. For example, when he treated worms with the fixative formaldehyde, cells that made the neurotransmitter dopamine acquired a phosphorescent glow, making them stand out against the worm's black background like a lantern on a dark street. Other treatments and stains lit up other types of cells.

As Sulston began constructing this cellular directory, it occurred to him that he might not need to limit himself to predefined classes of cells. He knew from his earlier studies that the adult worm contained roughly a thousand cells, each of which arose from a single zygote. All things being equal, a cell would need to divide only 10 or so times to generate a thousand cells ($2^{10} = 1{,}024$). Sulston imagined that with sufficient patience, it might be possible to track all the successive divisions of that founding cell. Doing so would result in a map indicating whence every cell in the adult worm had descended.

Cellular ancestries, like their human counterparts, are known as "lineages," and the notion of following them over time is known as "lineage tracing." The concept was not new—as far back as the late nineteenth century, naturalists like Edwin Conklin had followed the fates of dividing cells during the early stages of invertebrate development. But no one had conceived of mapping the lineage of an *entire organism*, a task that would be the cellular equivalent of naming every descendant of someone born 400 years earlier. Sulston's advantage was that he needn't rely on historical records. Since worm development took place over the span of a few days, he could simply watch cells through successive generations, seeing what each one did and what became of its offspring.

And watch he did, starting with the worm's larva. Together with Bob Horvitz, a mathematician turned molecular biologist who was a new addition to the worm group, Sulston sat at the microscope hour after hour, watching cells divide and differentiate into recognizable shapes. So painstaking were his observations of what each cell did, and when, that the wheels on his chair slowly carved a groove in the cement underneath. A friendship developed between Sulston and Horvitz, and over the next several months they mapped the lineage of the *C. elegans* larva, meticulously watching each cell make its way from adolescence to adulthood.

The effort yielded three remarkable insights. First, the total number of cells in the *C. elegans* body turned out, incredibly, to be the same from one worm to the next. When they hatched from their protective eggshells, every larva comprised 558 cells, while every adult comprised 959. The worm, it seemed, had little tolerance for deviation.

The second conclusion also related to invariance. Not only was the total number of cells conserved from one animal to the next—how they got there was identical as well. A given cell at a given position in one embryo behaved precisely the same as a cell in the corresponding position of another embryo. Nature's method for building an animal—at least one as simple as the nematode—appeared to be reproducible and exacting, its methods of construction preserved from one individual to the next.

The third discovery concerned cell loss, and in some ways, it was the most surprising and important finding of all. Cell death was a well-known feature of adult and embryonic tissues across the animal kingdom, and it was presumed to be a passive result of wear and tear. But the lineage diagrams demonstrated that this was not true. Sulston observed that it was always the *same cells* that perished, and they always did so at the same time, across the many individual animals he examined. There was nothing random about this cellular demise; instead, it was *programmed.* The invariant patterns of cell proliferation and differentiation that characterized the *C. elegans* lineage were accompanied by another uniform outcome: cell mortality.

Having defined the worm's cellular ancestry from larva to adult, Sulston set his sights on the earlier stages of development, tracing the lineage all the way back to its unicellular starting point. This was even more challenging, requiring the help of several newly recruited colleagues. But after months of intense observation, the team succeeded in completing the lineage, charting the worm's progress from one cell (denoted "P0") to two cells ("AB" and "P1") to four cells ("ABa," "ABp," "EMS," and "P2") and so on, following each successive generation until development was complete.

Again, the cellular trajectories were invariant from one animal to the next—hardwired and unchanging. In other words, if you were to

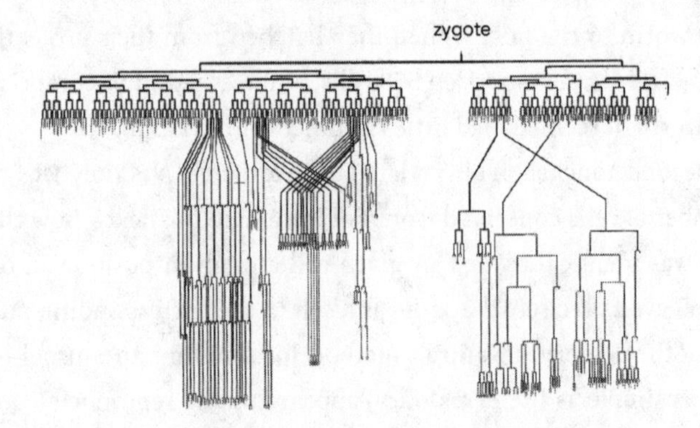

The complete lineage of *C. elegans*, from zygote to full-grown animal. The ancestry of every adult cell (the terminal point of each line on the diagram) can be traced back to the embryo's first cellular division.

point to any of the 959 cells in an adult worm, John Sulston could tell you how it got there, one cell division at a time. For as long as *C. elegans* exists as a species, its lineage—the animal's path to adulthood from one cell—will be the same.

FROM MUTANT TO FUNCTION

WE ALREADY NOTED THAT THE BEAUTY OF THE GENETIC APPROACH is that with the right tools, in the right hands, it can reveal the How of development. But prior to the late 1970s, when recombinant DNA technology arrived on the scene, moving from mutant to function was a near-insurmountable stumbling block. It was one thing to do screens and map the corresponding genes to their chromosomal homes. But retrieving the physical bits of DNA responsible for mutant phenotypes, and sequencing them, was another matter. Without a means of navigating the millions or billions of G's, A's, T's, and C's that make up an animal's genome, geneticists had no easy way to figure out which gene was responsible for a given mutant phenotype.

The methods of recombinant DNA—the ability to isolate genes physically, manipulate them, and determine their sequences—changed everything. It was the missing toolbox, and it gave researchers long-awaited access to the genetic material responsible for the mutations they had discovered—the molecular typos that manifested themselves as abnormalities in shape, movement, or lineage. Studies of development, which before had consisted of logging animal oddities, became an intensive exercise in reverse engineering: from mutant to gene, from gene to DNA sequence, from DNA sequence to protein sequence, and from protein sequence to protein function. The age of How had finally arrived.

Each of the mutants to emerge from the Heidelberg screen was a potential treasure trove of information. But molecularly characterizing each responsible gene—deducing its DNA sequence and determining the molecular function of the protein it encoded—was far too much for a single individual or even an entire laboratory to handle. As the Heidelberg scientists went their separate ways—Nüsslein-Volhard to Tübingen and Wieschaus to Princeton—they populated their new labs with a fresh generation of graduate students and postdoctoral fellows, giving each of them a small subset of mutants to work on. These apprentices, with time, started their own laboratories, forming a scientific lineage that lives on today.

As the data from this cadre of new scientists rolled in, the proteins encoded by the Heidelberg screen began falling into categories, molecular baskets of activity shaping the body plan of the fly. Transcription factors—the DNA-binding proteins that serve to regulate the expression of other genes—constituted one of the larger and more interesting categories. The fact that so many of the genetic mediators of shape were responsible for regulating other genes suggested a certain economy to development, an information cascade that was set in motion by a small set of instructions that then fanned out to influence the entire genome.

As these transcription factor genes, and their relatives, came under greater scrutiny, some were found to encode a unique protein domain, roughly 70 amino acids long, that was absent from other transcription factors. Disruption of these genes resulted in homeotic transformations, the substitution of body parts as in the four-winged fly we saw earlier.

Consequently, the DNA sequence encoding this protein domain came to be known as a "homeobox." Later, it became apparent that homeobox-containing genes exhibited another unique property: within the *Drosophila* genome, they line up, one after another, in a neat gene array, their chromosomal order mirroring the body parts whose shapes they were responsible for forming.

Other genes encoded proteins with different functions. Some were secreted, the mediators of cell-cell communication, while others were confined to the cell's cytoplasmic interior, serving as carriers of signals from the cell surface to its nucleus. But the most remarkable finding to emerge from all of this sequencing—a fundamental truth that would henceforth change our understanding of development *and* evolution—came about as DNA sequences from other organisms were compared to those of the fly. With almost no exceptions, all the genes identified in the Heidelberg screen were found to share counterparts in every animal examined, from fish to fowl. We know the genes are related because the similarities in nucleotide and amino acid sequences are far too close to have occurred by chance. Even homeobox-containing genes (or "Hox" genes), with their unusual chromosomal configurations, have such relatives, or "orthologs," that maintain the same ordered configurations along their chromosomes. This similarity of gene sequence, and presumed protein function, extends all the way to human beings. It is a principle known as "functional conservation."

As scientists have compared mechanisms of development across evolution—a field with the cryptic name "Evo-Devo"—it has become clear that nature reuses and reproduces the same cache of genes to create new branches on the tree of life. Genes involved in early development, including those identified in the Heidelberg screen, are particularly prone to duplication. For example, the fly genome holds only 8 Hox genes, while the mammalian genome holds nearly 40.

Hence, the genes that Wieschaus and Nüsslein-Volhard identified are the blueprints for the body's master plan not merely in *Drosophila* but across the entire animal kingdom. Indeed, one study comparing the genomes of flies and humans found that at least 75 percent of the genes that cause human diseases have counterparts in *Drosophila*. Some act by guiding cellular conversations, their names—*hedgehog, notch, wingless*, and *armadillo*—reflecting the misshapen bodies that emerge when they are mutated. Others, like the homeobox-containing *ultrabithorax* and *abdominal B* genes, regulate the expression of other genes.

It appears that nature builds a human being in much the same way that it builds a fly, despite the 600 million years of evolution that separate the two.

A SIMILAR STORY PLAYED OUT IN *C. ELEGANS*.

Bob Horvitz, who had been instrumental to Sulston's lineage-tracing efforts, recognized that the genetic approach could likewise identify genes that control cell fate. Reasoning that such genes might reveal an additional layer of evolutionary conservation, Horvitz proposed a new screen. As we saw earlier, Sydney Brenner had worked out the methods for performing mutagenesis in *C. elegans* a decade beforehand. Now, Horvitz would use the same approach to find mutants whose lineage, rather than shape or movement, was perturbed—mutants in which cell fates diverged from the lineage patterns he and Sulston had so carefully mapped out, appearing earlier or later than they should have, or never appearing at all.

Sulston agreed, and together the two scientists plotted a strategy. They would begin with the worm's reproductive system, as its cellular makeup—muscle cells, vulval cells, and neurons—was easy to recognize. Moreover, the absence of any of these cell types would result in an easily scored phenotype: an inability to lay eggs. Their initial effort yielded 24 such mutants, and it was proof that the genetic approach could also be used to find genes affecting a cell's lineage.

In 1978 Horvitz moved from Cambridge, England, to Cambridge, Massachusetts, keen to decipher the functions of these mutant genes in his new laboratory at MIT. (Sulston's interests, meanwhile, had returned to the worm's genome, a focus that later led to the sequencing of *C. elegans*. It was the first animal to have its genome completely sequenced.) Inspired by his earlier success, Horvitz also designed new screens to identify genes acting at much earlier stages of development, isolating hundreds of mutants affecting various worm lineages over the next decade and a half. The genes that he and his scientific progeny found encoded a wide variety of proteins—some released by cells, some remaining in the cytoplasm, and a smattering of transcription factors—a repertoire similar to the Heidelberg screen. Interestingly, some worm genes were found not to encode proteins at all. Instead, it was the "non-coding" RNA products of these genes—RNA molecules that bore no messages—that were responsible for regulating the expression of other genes.

But the most interesting mutants, and the ones that Horvitz's Nobel Prize citation would recognize, were those affecting cell death. Rather than an absence of cells that should have been present, these mutant animals contained cells that shouldn't have been at all. Sulston and Horvitz's lineage studies had revealed cell death to be a normal, deliberate process. Specifically, it was the programmed destiny of 131 cells born during *C. elegans* development to never make it to adulthood. Horvitz's screens identified the genetic basis for this "intentional" dance of death—genes whose job is to kill cells or to protect them from being killed. As with the genes identified in the Heidelberg screen, the worm cell-death genes have counterparts across the animal kingdom, where their death-inducing or death-preventing activities ensure that development proceeds normally.

With time, Horvitz's students and other researchers studying these genes would find that programmed cell death, or "apoptosis," is not limited to embryos; adult tissues use the same programs to maintain their equilibrium. When the immune system wishes to eliminate cells infected by viruses, for example, apoptosis is a favored means of elimination. Conversely, when other parts of the body come under stress, *anti*-apoptotic programs step in to protect cells from dying. The regulation of cell death, either positively or negatively, is ubiquitous across the plant and animal worlds.

BY THE 1990S, THE FIELDS OF DEVELOPMENTAL BIOLOGY, GENETICS, molecular biology, and evolutionary biology had essentially merged. The scientists populating these disciplines used different tools, but they all had something to add regarding the One Cell Problem. And by applying the genetic approach to worms and flies, they had arrived at a partial answer: a finite number of gene products—transcription factors, soluble proteins, intracellular signaling molecules, and non-coding RNAs—bear the responsibility for building a body, whether that body belongs to a fish, a dinosaur, or an orangutan. Nature, in its thrift, recycles the same design principles repeatedly during development, duplicating or editing genetic programs that predate even *C. elegans* and *D. melanogaster.*

The physicist Richard Feynman once said, "What I cannot create, I do not understand." It is a high bar for measuring knowledge, and one that developmental biologists remain far from achieving. Between the many genetic screens that researchers have performed in worms, flies, and other species, we have discovered most if not all of the genes with a universal role in development, and we have a rough idea of how their protein products work. Consequently, it is not regarding the identity and function of these individual genes, and their products, where our comprehension falls short. Rather, it is the bigger picture of development—how genes work together to build the interactive networks from which our bodily cities and cellular societies rise—that remains elusive. In that sense, we have likely only skimmed the surface of the How. French scientist-philosopher Jean Rostand put it best, noting: "Biologists come, and biologists go. The frog remains."

Chapter 6

DIRECTIONS, PLEASE!

Form ever follows function, and this is the law.

—LOUIS SULLIVAN, "THE TALL OFFICE BUILDING ARTISTICALLY CONSIDERED"

We live in a three-dimensional world. All objects, living or not, have a top and a bottom, a front and a back, a left and a right. Within this space, the potential for new forms is nearly endless, and evolution has taken full advantage. Cells come in many shapes and sizes, and they can spread out to form flat tissue planes, round up into tubes, venture into unpopulated areas, or seek out high-density neighborhoods. A cell may stretch out its limbs or withdraw them, squeeze, expand, move in waves, or stand its ground. These attributes, in aggregate, are what give rise to bodily form, and it is form—an animal's size, proportions, and mobility—that determines whether it will live a long, healthy life or end up as someone else's lunch.

Early in embryonic life, when we are still a mere ball of cells, standard dimensions do not exist. The embryo must work its way into our three-dimensional world, positioning its organs correctly and scaling them accordingly. Shapes emerge sequentially, each addition dependent on one that preceded it. Relationships are everything, for a cell's position in the embryo is not specified by *x*, *y*, and *z* coordinates, the way an object might be rendered in a computer design program. Rather, cells "know" their places solely in relation to other cells.

As we saw earlier, the embryo is home to massive amounts of cellular cross talk, where cells communicate their location or other bits of information through molecular signaling. But it is not merely the presence or absence of a signal that gives a cell its footing; the *level* of signaling also matters. When a cell transmits a molecular message—a secreted protein or other information-rich molecule—the molecule's concentration drops off the farther it gets from the source. Nearby cells receive a "strong" signal, while cells that are farther away receive a "weak" one, and this makes a difference. Just as you can tell how far you are from a cellular tower based on the number of bars on your mobile phone, cells in the body determine their distance from the source of a signal based on molecular concentrations. Such differences in signaling intensity are often known as "gradients," and the molecules that act as proximity indicators are called "morphogens." In addition to such "soluble factors"—molecules that can diffuse away from the source—cells can also be shaped via direct contact or through physical forces, which push or pull cells in one direction or another.

Until now, our consideration of the One Cell Problem has focused on differentiation, the instructions that impart cellular identity. But this leaves a second and equally important question unanswered: How do cells know not just what to become but where to go? That part of the embryo's destiny—the molding of tissues into their mature shapes—is realized in partnership with differentiation, as patterning programs spelled out early in development are fulfilled. This sculpting process is known as *morphogenesis*, and it is the fascinating yet mysterious science of how the embryo achieves its three-dimensional character.

THE BIOLOGIST LEWIS WOLPERT ONCE SAID, "IT IS NOT BIRTH, MARriage, or death, but gastrulation which is truly the most important time in your life." Obviously, this pronouncement was tongue-in-cheek. But it highlights gastrulation as a key milestone in development—the point at which undifferentiated cells begin to receive their tissue assignments en masse. During this process, plasticity, or the ability of a cell to change its identity, takes a nosedive, as Briggs, King, and Gurdon observed in their nuclear transplantation experiments. But in parallel with these changes in cell fate, the cells of the embryo spatially reorient themselves in dramatic fashion, taking their first step toward a mature form.

In mammals, fertilization takes place in the oviduct, or fallopian tube. Only a single sperm is permitted to enter the egg, and the result is a zygote bearing two copies of each chromosome—what scientists call a *diploid* embryo. The zygote then begins its cleavage divisions, and once it has divided more than two or three times, it is considered a *morula* (from the Latin *morum*, meaning "mulberry"), a sphere roughly 8 to 30 cells in size. Barely visible to the naked eye, the embryo works its way down the oviduct until it reaches the uterus. There, the endometrial cells lining the uterine wall are ready and waiting, conditioned by the mother's hormones to receive the new embryo. The following step—implantation—embeds the embryo safely in this maternal incubator, where it lives until it is born.

By now, the embryo has matured into a blastocyst, a hollowed-out sphere of several dozen cells that is the mammalian equivalent of a blastula. Within the blastocyst, two types of cells are distinguishable. One population is the *inner cell mass*, an asymmetrically located cluster of cells that develop into the fetus and future animal. The other population is the *trophectoderm*, which makes up the blastocyst's outer shell and forms the placenta. It is the derivatives of the trophectoderm that

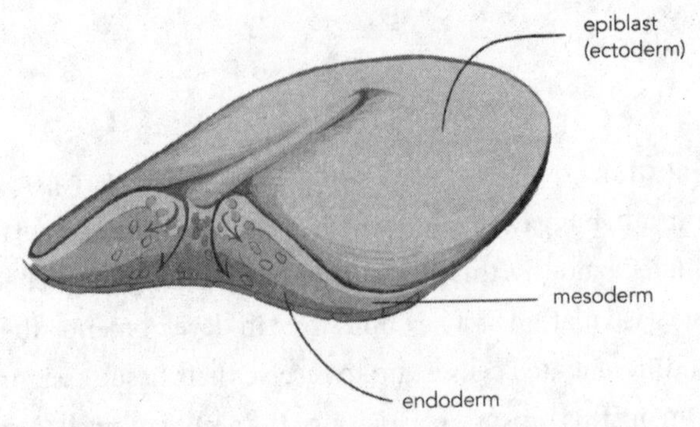

During gastrulation, epiblast cells migrate through the primitive streak (central groove) to give rise to the three germ layers. Cells on the bottommost layer form the endoderm, cells that remain on the upper layer form the ectoderm, and those in between form the mesoderm.

make implantation possible and that will later provide nutrition from the mother to the fetus.

As the embryo prepares for gastrulation, cells within the inner cell mass flatten into a single-layered sheet of cells called the *epiblast*. An indentation appears on the epiblast surface and expands, growing into a trench called the "primitive streak." Cells surrounding the fissure are swallowed up by the dozens, like celestial bodies drawn to a black hole, and they emerge on the other side transformed. Gastrulation is a developmental rite of passage—a Harry Potteresque "sorting hat" of development—and once cells pass through the primitive streak, they now belong to one of three embryonic *germ layers*: the *ectoderm*, the *mesoderm*, or the *endoderm*.

We can think of germ layers as academies—the embryonic equivalents of a four-year college, a vocational school, and a seminary. A college graduate is more likely to become an entrepreneur than a chaplain; a vocational-school graduate is more likely to become a technician than a professor. Likewise, the cells belonging to each germ layer tend to favor certain paths over others. But just as the most vibrant societies are those whose citizens come from diverse backgrounds, most of our tissues comprise cells from all three germ layers. It is the job of morphogenesis to take diverse cellular elements—the new graduates of gastrulation—and weave them together into coherent, functioning units.

UP, DOWN, OUTSIDE, INSIDE

IN EMBRYOS AND ADULT ANIMALS, CELLS COME IN TWO MAIN FLAVORS: *epithelial* and *mesenchymal*. Most epithelial cells are derived from the endoderm or ectoderm germ layers, and they are defined by their ability to adhere tightly to one another, forming barriers. The keratinocyte layer of the skin, comprising the body's protective surface, is the most visible example of an epithelium. But the body's internal organs also play host to epithelia, whose cells line the tubes connecting our internal organs to the outside world. Mesenchymal cells—the stuff of muscles, bone, tendons, and cartilage—lack such outward-facing surfaces, but their role is no less important, for they constitute the glue holding the body together and the engine driving its actions.

Most organs are derived from two, or perhaps all three, germ layers, and they contain a mixture of epithelial cells and mesenchymal cells. Take the lung, for example, with its endoderm-derived airways. These conduits, which transport oxygen in and carbon dioxide out, are lined by epithelial cells but are surrounded by mesoderm-derived muscles and ectoderm-derived nerves. Developmental biologists, in casual conversation, may refer to an organ as originating from a single germ layer—calling the lung an "endoderm derivative," for example. But this is merely a shorthand for the tissue's most important cellular component; in the lung, this honor belongs to the endoderm-derived epithelial cells, which exchange life-sustaining oxygen for carbon dioxide. Other organs receive different germ layer attributions: for example, the heart and kidneys might be referred to as "mesenchymal derivatives" while the brain and skin might be dubbed "ectodermal derivatives." But of course, the reality is far more complex.

The defining feature of an epithelial cell is its tight adherence to its neighbors. Such linkages are mediated by *adhesion complexes*, clusters of different proteins that prevent almost everything—lipids, amino acids, ions, and even water—from passing between cells. Consequently, the only way for material to get across an epithelial barrier is go *through* cells rather than around them. This active process is facilitated by transporters and channels, molecular gateways whose job it is to pick and choose what substances can go across. To do this, epithelial cells must have the ability to distinguish top from bottom—the "apical" and "basal" surfaces of the cells, respectively. In the gut, for example, the apical membrane of an intestinal epithelial cell faces the internal lining of the tube where the digested products of a meal are located. Dietary nutrients are taken up through channels on this apical surface and transported (across the cytoplasm) to the basal surface, where they are then delivered to the circulation. Only those substances that are properly broken down can pass through this epithelial barrier. Epithelial cells are the body's customs officers, permitting only authorized objects to enter.

HOW DO CELLS FIND THEIR WAY IN THE DEVELOPING EMBRYO? WHAT draws epiblast cells toward the primitive streak and then, after gastrulation, directs them to their final destinations? How do intricate tissue structures, the airways of the lung or the electrical system of the heart, achieve their brilliant configurations, and how do neurons know to stretch their axons this way instead of that to form the proper *synapses*? In short, what drives the sculpting of form?

It should come as no surprise that tissues are shaped through cell-cell communication. We have already seen some dramatic examples, such as the phenomenon of induction demonstrated by Spemann and Mangold's discovery of the embryonic organizer. In that case, cells in one region of the embryo (the "dorsal lip") dispatched a message leading to the formation of an unexpected twin. But induction need not be so dramatic; most developmental communiqués are routine and have more modest consequences. Nevertheless, such molecular conversations are the basis for virtually all of physiology, from wound healing to the quickening of a heartbeat to the realization that it is time for another cup of coffee.

Whereas Spemann and Mangold could only imagine the basis of cellular cross talk, we now understand that these ubiquitous conversations are mediated by "signaling pathways" like those we have seen before—evolutionarily conserved molecular cascades that regulate the shape and pattern of the early embryo. Some pathways require both parties (cells) to be immediately adjacent to each other, as if speaking in person. Others operate across a somewhat greater space, forming gradients in which a signal's strength, not merely its presence or absence, conveys information. Finally, signals operating across vast distance—such as those initiated by hormones like insulin or growth hormone—circulate throughout the body via the bloodstream, giving them an unlimited range of activity.

If we think of these modes of communication as various forms of broadcast media—newspapers, TV, telephone, or radio—then what Marshall McLuhan said of society is also true of the embryo: "The medium is the message." Each flavor of molecular signal carries a different meaning for the cell, and while we might expect hundreds of signaling molecules to be involved in providing cells with instructions, embryos rely on only a dozen or so distinct systems of communication.

Context is equally important. A cell's ability to react to a signal, and the specifics of its response, are known as *competence*. Given an identical developmental signal, a mesodermal cell in the kidney and an ectodermal cell in the spinal cord may derive completely different meanings. This is because a cell's *epigenetic* makeup—the unique identifiers that brand each cell's genome differently—defines the cellular response to a given signal. (Picture how following a teacher's instructions to "turn your textbook to page 65" would have very different consequences for a student in a health class versus a student in calculus. We will have more to say about epigenetics later.)

Finally, embryos have other ways to enhance the diversity and complexity of signaling. One method is to employ combinations. For example, a cell's response to signal A can differ depending on whether that signal is received alone or accompanied by signal B. Another near-universal feature of cellular cross talk is reciprocal signaling, a form of bidirectional communication by which cells reply to the messages they receive, resulting in a fine-tuned back-and-forth that transpires with temporal and spatial precision. And because the interpretation of these molecular memoranda also depends on location—a cell's position in three-dimensional space—embryonic cells can calculate, moment by moment, where they are and where they need to be.

ONE EXAMPLE OF THIS INTERPLAY IS THE VERY FIRST CELLULAR "decision" taken in a mammalian embryo, which occurs well before gastrulation. The transition of an embryo from morula to blastocyst, once the embryonic cells have cleaved roughly five times, occurs when a cavity called the blastocoel (pronounced "blast-o-seal") forms on one side of the morula cluster. The resulting lopsided embryo is populated by two types of cells—the inner cell mass (ICM) and the trophectoderm (TE)—

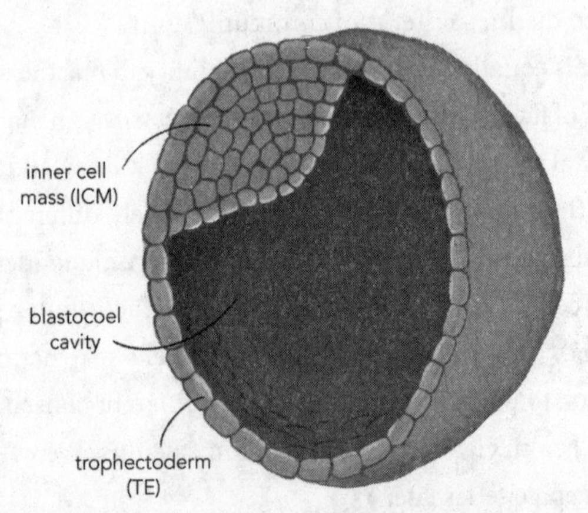

The mammalian blastocyst contains approximately 100 cells surrounding a fluid-filled blastocoel cavity. Cells residing within the inner cell mass will give rise to the embryo and future animal, while those in the trophectoderm will give rise to the placenta.

whose descendants have vastly different fortunes. (As noted earlier, the derivatives of the ICM have the privilege of giving rise to the future animal, while those of the TE, vital for the placenta during pregnancy, are discarded following delivery.)

For decades, the molecular basis, and timing, of the momentous ICM/TE "choice" has been a central question in developmental biology circles. In the late 1960s, Polish biologists Andrzej Tarkowski and Joanna Wróblewska proposed what they termed the "inside-outside" model of early mammalian development. It postulated that a cell's position in the morula—whether it was found in the interior of the grape-like cluster or its periphery—ordained the fate of its descendants. According to the model, cells located on the inside of a morula would form the ICM, while those on the outside would form the TE.

Nearly two decades later, mouse embryologist Roger Pedersen confirmed this association by following the fate of individual morula cells, a lineage-tracing approach akin to the one John Sulston and his colleagues used in their studies of *C. elegans* development. Pedersen found that "inner" cells did not exist before the 16-cell stage of embryogenesis (at the 8-cell stage, cells are arranged so that all cells face the "outside"). Using a dye to label the outer cells at the 16-cell stage, Pederson found that these cells had a strong, but not absolute, tendency to become TE. As time passed and the embryo matured, this propensity strengthened, and eventually a cell's position—whether inner or outer—became irreversibly tied to its fate.

These observations raised two questions: First, how does a cell "know" (at the molecular level) whether it is on the inside or the outside of a cluster? Second, how does its fate go from being probabilistic to irrevocable over time?

To consider the first question, pretend for a moment that you are one of a dozen or so cells in an early-stage morula. If you are positioned in the cluster's interior, that means you are surrounded by other cells, as if standing in the middle of a crowd. But if you reside on the cluster's outskirts, there are cells on one side of you but not the other; part of your body is flanked by a barricade, a protective layer called the "zona pellucida" that serves as the equivalent of a mammalian eggshell.

This topological difference has molecular consequences, conveyed by an evolutionarily ancient signaling system known as the "Hippo pathway." Every time a cell touches another cell, receptors on the cell surface activate the Hippo signaling pathway. The resulting molecular relay makes its way to the nucleus, prompting changes in the activity of fate-determining transcription factors. When the level of Hippo signaling is high, as occurs in the morula's interior, a transcription factor known as OCT4 is expressed. When the Hippo signal is lower, a transcription factor known as CDX2 is expressed. In turn, these two transcription factors regulate hundreds of other genes that collectively determine whether a cell becomes ICM or TE. OCT4 and CDX2 are thus considered "master regulators" of cell fate.

The second question concerns the increased odds of an outer cell becoming TE (and an inner cell becoming ICM) as the embryo matures. Here again, the answer is a phenomenon we have encountered before—the principle of cellular plasticity, which tends to decrease as development proceeds and cells become committed to their destinies. In the context of the blastocyst, we have a molecular explanation for the phenomenon: positive and negative feedback loops. As it happens, OCT4 and CDX2 transcription factors—the master regulators of the ICM and TE fates—also regulate each other. Specifically, OCT4 activates its own expression and represses that of CDX2. Conversely, CDX2 promotes its own expression while repressing that of OCT4. In practical terms, this means that embryonic cells that are "tipped" in one direction or another—toward ICM or toward TE—become progressively more "locked in" as time goes on. These molecular details are specific to blastocyst development, but similar feedback loops are ubiquitous in embryonic development and differentiation, and they help resolve the paradox between a cell's potential (the many roles it *could* adopt) and its final identity (the one role it *commits* to).

But this raises yet another chicken-and-egg question: What came before? Does an "inside" cell in the center of a morula find itself there by chance, or is there another, earlier signal that told it where to go? While the inside-out model partially explains how cells become part of the ICM or TE, we will soon see that other factors influence cell fate beforehand, possibly as early as the zygote's second or third cleavage. It makes one wonder whether our cells ever have free will, whether their destinies were ever truly unrestricted. Or are they—like the cells of the sea squirt or the developing worm—merely players in a drama whose ending is all too predictable?

THINKING IN OTHER DIRECTIONS

IF YOU HAVE EVER TRIED TO PET A CAT FROM HEAD TO HAUNCH, AND then reversed direction (likely evoking an unpleasant response), then you have come face-to-face with another feature of morphogenesis called *planar cell polarity.* Earlier, we learned that epithelial cells can distinguish their tops from their bottoms, but their proprioceptive abilities do not end there; epithelial cells can also tell the difference between front and back. Beyond ensuring that hair shafts in the skin point the same way, planar cell polarity ensures (among other things) that the intestines propel food in the right direction and that the hair cells of the inner ear are properly aligned, giving us our sense of balance.

If cells can tell front from back, and top from bottom, can they also tell left from right? It depends. Many animals are completely symmetrical, with no difference between left and right. But other creatures, including human beings, are "lateralized"—such that certain organs like the liver are on the right side of the body while other organs like the spleen and heart are on the left. This asymmetry arises early in development, around the time of gastrulation, because of specialized cells in a region near the primitive streak known as the "node." These cells have propeller-like structures on their surface called "cilia" that can spin in only one direction: clockwise. This unidirectional rotation of the cilia causes fluid sitting above the epiblast layer—which is rich with signaling molecules—

to flow from right to left. This, in turn, creates a gradient of secondary signals that round out this critical third dimension of the body.

BEYOND THE STANDARD THREE DIMENSIONS THAT WE EXPERIENCE IN our daily lives, the embryo has other spatial relationships, or axes, to worry about. One example is the *radial* axis—the differences in form and function that depend on a cell's position relative to the center of a tube. The intestine is a good example, with its cells arranged in distinguishable layers as one moves farther away from the gut's hollow interior. The innermost layer is the mucosa—the endoderm-derived epithelial cells that absorb dietary nutrients and secrete lubricating fluid into the intestinal space. Beneath the mucosa is the submucosa, a mixture of mesoderm-derived blood vessels and connective tissue that swiftly incorporates newly absorbed nutrients into the bloodstream and distributes them to the rest of the body. Below the submucosa are additional layers of muscle and nerve whose activities are coordinated to propel food (in only one direction, thanks to planar cell polarity) from the esophagus to the colon.

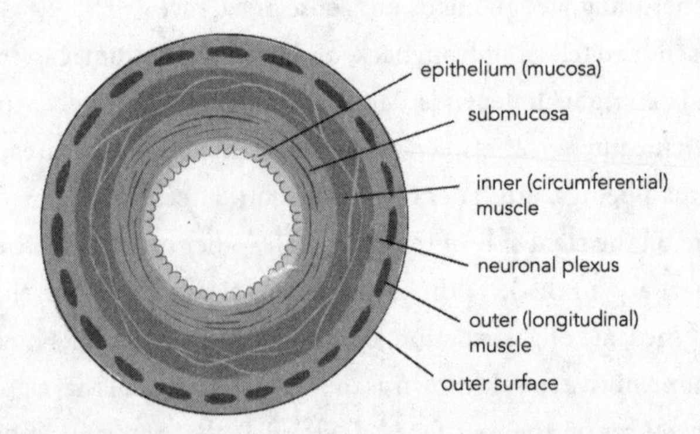

Tubular organs, like the gut, have a radial axis. A cell's identity, and its function, depends on its position relative to the tube's center and its outermost layer.

The ordering of the intestinal layers does not occur by chance. Rather, a signaling molecule released by the epithelial cells of the nascent gut tube diffuses to the surrounding (undifferentiated) mesenchymal cells. The molecule in question is a member of the hedgehog family of proteins, whose founding member was discovered in the *Drosophila* Heidelberg screen. Earlier, we compared the distance from a cell tower and the corresponding strength of a cell-phone signal. Similarly, mesenchymal cells "see" different levels of hedgehog signaling depending on how far away they are from the center of the tube. Those closest to the intestinal interior, receiving the "highest" level of hedgehog signaling, give rise to the submucosa, while those farther away receive lower levels of signaling and become muscle instead. Having received these differentiating signals from the epithelium, the mesenchymal cells return the favor; they release a different signaling molecule that circles back to the epithelial cells, telling them to proceed with their own maturation programs.

This type of communication is called epithelial-mesenchymal signaling, and it is also relevant for tissue formation along another axis: the proximal-distal axis, which distinguishes cells that are close to the body's center from ones that located are farther away. The upper limb is a great example of this kind of form-building. If we follow the arm from the proximal shoulder to the most distal fingertips, we see that each part of the upper limb has a different structure and function. The shoulder and upper arm are endowed with strength and range of motion, enabling the arm to lift, push, swat, or wave. By contrast, the hands and fingers are built for manipulations on a much finer scale, like typing this sentence or performing microsurgery on a frog embryo. Similar differences demarcate different regions of the lower limbs and feet.

Like the gut tube, the arm owes its shape to the relative levels of signaling that its precursor cells receive. While organs are forming inside the body, the arms and legs begin as "limb buds," tiny clusters of cells on the embryo's surface. The limb bud starts out with two components: a cluster of undifferentiated mesenchymal cells lying beneath a layer of ectoderm-derived epithelial cells. The epithelium begins the dialogue, and the mesenchyme responds as it does in the intestine, its cells recognizing their positions in space and time based on the strength of the epithelial signal. As the exchange continues, other cells join in, taking their conversations to different corners of the developing appendages. The talk turns to action—growth and differentiation—as each domain of each extremity recognizes its calling. The proximal part of the upper limb builds muscle and assembles the shoulder girdle, while the distal part works on forming the many bones of the wrist, hand, and fingers. Nerves are invited to participate. A similar process plays out in the lower limb. By the time the arms and legs have formed, the conversations between epithelium and mesenchyme, which initially sparked the process, have long been forgotten.

TO GET SOMEONE TO CHANGE THEIR POSITION FROM POINT A TO point B, you can attempt one of two things: convince them to move of their own accord or attempt to physically reposition them. Morphogenesis relies on signaling pathways like Hippo and hedgehog to guide cellular movements, but ultimately it is the application of force—a push, pull, twist, or flow—that shapes tissues. Without force, either internally generated or externally applied, a cell will not move.

In the simplest case, cells may reach out with a small part of their membranes and latch on to whatever surface they find. In the presence of other signals, they may then engage an intracellular pulley system—the *cytoskeleton*—with which cells can drag themselves toward the point of attachment. Through this repeated process of reaching out and drawing in, like that of a climber scaling a wall, a single cell can travel great distances.

In other circumstances, cells move collectively, like an audience filtering back into a theater at the end of intermission. Such group movements are common during development, particularly when a tissue needs to get longer but not wider. Embryos have several ways to achieve this, and one common method is a phenomenon called *convergent exten-*

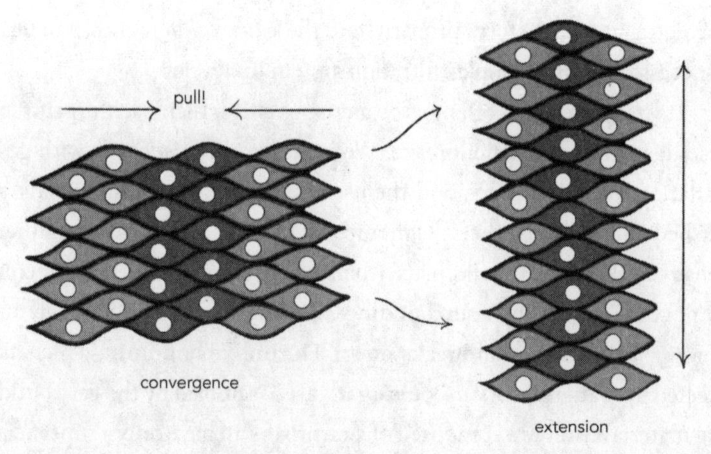

During convergent extension, cells apply tension to their neighbors, pulling them toward the center. As the middle becomes more crowded, cells run out of room, causing the tissue to elongate perpendicular to the axis of force. This allows certain parts of the body to become longer as development proceeds.

sion. To understand the process, picture a crowd of people distributed on either side of an imaginary line. Now, envision a series of ropes connecting those on one side of the line with those on the other side. On command, everyone begins to pull, bringing them closer to the imaginary center line. But as the two sides *converge* on each other, there is only so much room in the center; hence, the crowd is forced to "extend" itself along the perpendicular axis.

These examples show how *internal* forces shape the embryo, as cells pull against each other or crawl along surfaces to mold tissues. But *external* forces also play a role. One example, discovered by Harvard biologist Cliff Tabin, has to do with the shaping of the intestines in chick embryos. Tabin's group found that the tissue surrounding the intestine—the mesentery—applies tension to the growing gut tube, causing it to loop back on itself. These forces continue to act as the intestine develops, leading to the formation of villi, the tiny projections that vastly increase the absorptive surface area of the intestinal epithelium. Fluids can also provide force. Examples include the heart, where blood flow helps the cardiac chambers to form properly, and the kidney, where transit of fluid is necessary for its complex filtration system to develop.

If we again think of embryos as rising cities, then each organ represents its own construction site. We have seen how, starting with gastrulation, our cells reposition themselves, drawn to certain signals or pushed and pulled by their neighbors. All the while, cells maintain their sense of self (identity) and place (their position in space) through constant contact with their surroundings. Beyond this point, however, our analogy to construction breaks down. For unlike buildings, which are erected by teams of workers, our organs are assembled by the very building materials they are composed of. Scientists still have only a superficial understanding of the molecular details of this process, but a few specific examples can give us a glimpse of how tissue self-assembly might work.

MAKING THINGS STICK

A CENTURY AGO, DEVELOPMENTAL BIOLOGISTS WERE DUMBFOUNDED by the embryo's ability to take corrective action, calling it a machine that seemed capable of assembling itself. We now know that cell-cell interactions, which tell a cell what and where it should be, are responsible for the embryo's regulative properties. The path followed by an individual cell depends on the history of signals it has received. Even *C. elegans*, with its seemingly invariant cellular lineage, is not immune to this conditionality; it is simply the reproducibility of the cellular responses to stimuli that gives worm development its appearance of invariance.

In addition to these signals, there is another force at work: differential adhesiveness—differences in the "affinities" cells have for each other—also plays a major role in shaping tissues. The process is driven by *adhesion molecules*, membrane proteins that act as a kind of Velcro for other cells. There are many families of adhesion molecules, and each has a unique set of affinities. When a cell encounters another cell bearing its favorite flavor of adhesion molecule, the cells are drawn to each other like the opposite poles of a magnet. (One such adhesion molecule—

E-cadherin—is the principal force driving the formation of impenetrable epithelial barriers, as E-cadherin molecules on neighboring cells grip each other tightly.) In the absence of such attractive forces, cells merely ignore each other.

The first demonstration that differential adhesion could, on its own, shape tissues was a 1955 study from Hans Holtfreter, a professor of biology at the University of Rochester, and his student Philip Townes. The researchers microdissected salamander embryos to create solutions in which all the cells floated independently from one another, then allowed the cells to reassociate. The cells aggregated as a large cluster. But when Holtfreter and Townes looked further, they found that there was an organization to the cellular mass: it consisted of three layers, with endoderm cells congregating at the bottom, mesoderm cells in the middle, and ectoderm cells on top. In other words, the cells belonging to these three

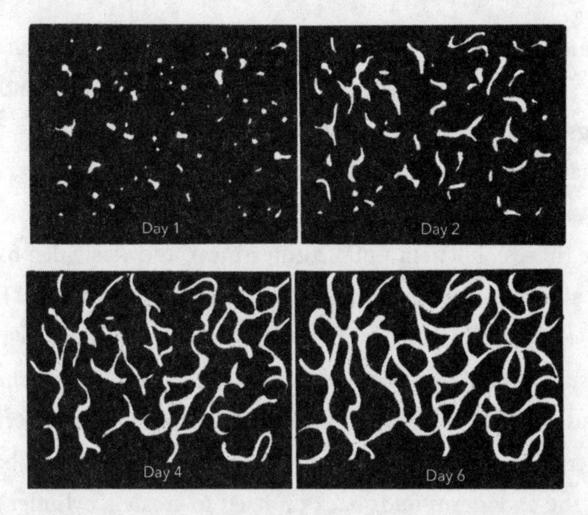

Cells have innate adhesive properties that allow them to associate and self-organize into structures. One example is how capillaries (small blood vessels) can self-assemble from dissociated vessel-forming endothelial cells. Shortly after being placed into a gelatin-like matrix, endothelial cells float freely and independently. But over several days, and without outside assistance, the cells find their neighbors and reconfigure themselves into hollow tubes that are nearly identical to vessels normally found in the body.

lineages are inherently drawn to other "like" cells, allowing a mixture of cells that had been pulled apart and widely dispersed to reconstitute the basic morphology of the post-gastrulation embryo.

This tendency for cells to associate with other cells in predictable ways is known as "self-organization," and its applications extend beyond studies of embryonic development to tissue engineering. One example of this is the self-organizing behavior of "endothelial cells," the cells that line our blood vessels. Much like epithelial cells, endothelial cells adhere tightly to one another, forming a barrier that keeps blood cells and plasma from crossing vessel walls. Remarkably, if you take a sample of isolated endothelial cells and suspend them in a gel, the cells find each other and form microscopic capillaries, like the patterns that form when magnetic tiles are allowed to self-associate.

GOING TUBULAR

LET'S SPEND A MOMENT THINKING ABOUT ONE SPECIFIC ANATOMICAL structure: the tube. We have already discussed the gut tube and its surroundings, but other hollow conduits—such as the capillaries, airways, and urinary tracts of animals—are equally crucial. Tubular networks increase surface area, expanding the number of cells available for gas exchange (lungs), nutrient uptake (intestines), and waste disposal (kidneys). In addition, tubes facilitate transportation, the principal function of blood vessels, the ventricles of the brain, and the bile ducts of the liver.

The simplest tube-forming program to understand is *branching morphogenesis*, a process akin to the branching of limbs from the trunk of a growing tree. In mammals, the first tube to develop is the gut tube, which forms as the endoderm folds back on itself to create a cylinder. Organs connected to this central tube—the lung, the liver, and the pancreas—can then bud off and develop independently. The lung is nature's finest example of branching morphogenesis, and it begins as a small outpouching from the gut tube near the future throat. This indentation, its scale comparable to a hole on the green of a golf course, appears when a small patch of mesoderm tells the overlying endoderm cells to begin to move below the epithelial plane. As the cellular pocket elongates, penetrating deep into the underlying tissue, new signals emerge on either side of the growing tube, causing it to branch. The process repeats, resulting in secondary branches, then tertiary branches, and so on. In humans, this translates into millions of gas-conducting airways and gas-exchanging alveoli which, if spread out, would have a surface area bigger than a badminton court.

Branching morphogenesis is responsible for the bifurcating structures of the lungs, liver, salivary glands, and other organs, but it is not the only way to make a tube. Nature employs other methods, such as the circular folding of an epithelium back on itself like a cigarette made from a square sheet of rolling paper. The neural tube, which contains the precursors of the nervous system, arises in this manner, and errors in folding result in "neural tube defects" such as spina bifida.

If a tube can form, it can also become blocked. While clogging is rare during embryogenesis, it is commonplace in adults. Heart attacks (blockage of coronary arteries), pancreatitis and cholangitis (blockage of the pancreatic or liver ducts), pulmonary embolism (blockage of the pulmonary artery), hydrocephalus (blockage of the brain's ventricles), and choking (blocked airways) are all consequences of obstruction. While physicians treat these obstructions by removing or bypassing the blockage, nature attempts its own remedy by making new tubes. In the circulatory system, this is called "collateralization," where new vascular channels allow blood to flow around an impassable major vessel. Here, as in other cases where the body creates a work-around, nature redeploys embryonic programs to solve a postnatal problem.

So far, we have considered the many structures the embryo must establish and draw out until they achieve their final forms—surfaces, layers, tubes, and extremities. These activities are coordinated with differentiation, as tissue shape and cellular identity arise in tandem. But there is a third element to the process of building a body, one that is far less understood than either differentiation or morphogenesis. This missing piece is the feature of size and proportionality, and without it our tissues would be unable to function.

SIZE MATTERS

ONE OF THE FIRST THINGS WE NOTICE WHEN WE COME FACE-TO-FACE with another creature is its size. What follows is a reflexive ancestral question: Can I eat it or can it eat me? Body mass in adult mammals ranges from about 1.5 grams in the Etruscan shrew to approximately 150,000,000 grams in the great blue whale, a hundred-million-fold difference. Yet all animals develop from eggs of similar size, regardless of their adult proportions. Instructions dictating organismal and organ size must lie deep within the genomes of all animals, and yet their identity is one of nature's best-kept secrets.

Although individual cells can vary a bit in size, these differences are not what define an animal's dimensions. Rather, the most important factor determining an organism's size, and that of its tissues, is cell number. This variable is controlled by several factors, including hormones that circulate through the body, like growth hormone and insulin-like growth factor (IGF). In humans, growth hormone deficiency leads to short stature, while in dogs, inherited variants in the IGF gene account for the diminutive stature of small breeds, including Chihuahuas, Pomeranians, and toy poodles. Nutrition also plays a role, as malnutrition can lead to growth defects either before or after birth.

Yet these factors do not fully explain how each animal species occupies only a small range within the shrew-to-whale spectrum. Biologist D'Arcy Wentworth Thompson pointed out: "We call a thing big or little with reference to what it is wont to be, as when we speak of a small elephant or a big rat." Nutrition fails to account for these differences, for the bodily dimensions of an overfed rat will still be those of a rat, even if it becomes obese. And while differences in hormones like IGF may account for differences in size within a species, they do not explain size variation across the animal kingdom. Rather, these differences are hardwired in the genetic code in ways we cannot yet see.

It is not only an organism's overall size that is noteworthy, but also the proportionality of its parts. In humans, an individual's arm span is generally equal to height. The liver almost always accounts for 2 percent of body weight. Our bodies are assembled symmetrically, such that the right arm differs in length from the left by less than half an inch. These ratios, while different from species to species, are remarkably consistent within a species. So how is such an accurate accounting of size, symmetry, and proportionality achieved?

As with so much in biology, the answer is likely a mixture of genes and environment. Among the most informative studies on the topic is a series of experiments on salamanders performed by Ross Harrison in the 1920s and extended a decade later by Victor Twitty and Joseph Schwind. The experiments were done by grafting limb buds, primordial tissues that would eventually grow into the limbs, between two species of the *Ambystoma* salamander, one larger (*A. tigranum*) and one smaller (*A. punctatum*). The experimental details varied, but in general, the grafted limb grew to the size it would have attained if left in its original position. In other words, the grafted cells "knew" how big the limb should be, even though the limb's development had barely begun. Harrison called this property the "growth potential" of the graft, and it reflected an *intrinsic* property of the transplanted cells. Nevertheless, other experiments revealed that *extrinsic* factors—such as the nutritional status of the host—also influenced the graft's development. The regulation of size is yet another example of nature's dance with nurture.

In my own work as a postdoctoral fellow, I found that a similar dichotomy between genetic programs and environmental influences governs the size of the pancreas and the liver. Applying genetic tricks, I was able to engineer mid-gestation mouse embryos in which either the pancreas or the liver was a third the normal size, and then allowed these embryos to develop to term. Surprisingly, the two organs—which emerge

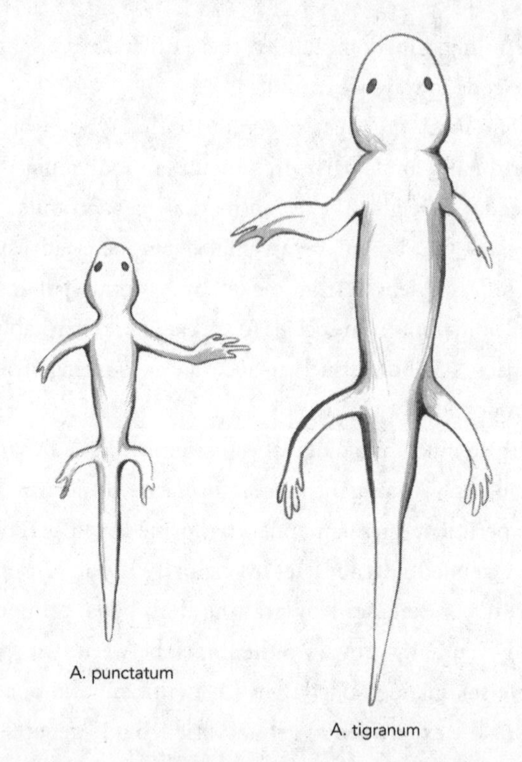

When limb buds from a small salamander species (*Ambystoma punctatum*) were reciprocally transplanted with those of a large species (*Ambystoma tigranum*), the size of the resulting limb tracked with the original source, not the recipient. This indicated that some aspects of tissue size are autonomously determined.

from patches in the endoderm immediately adjacent to each other—dealt with this embryological amputation in very different ways. The liver, capable of recognizing the size deficiency, compensated by increasing its growth rate and recovered its normal size by birth. The pancreas, by contrast, made no attempt to catch up; still only a third the size of a normal pancreas at birth, it remained small throughout the animal's life. These embryonic growth patterns foreshadow the ability of these organs to control growth later in life. If a surgeon removes part of the liver, the remaining portions will grow to recover the organ's preoperative size (we will return to this topic again later in the context of organ regeneration). The same kind of recovery does not occur after pancreas surgery.

An animal's size influences all aspects of its biology, including life span. More than 2,000 years ago, Aristotle noted the correlation between an organism's size and its longevity: larger animals live longer. We do not know why, but one theory links size and life span to metabolism. Because metabolic rate is inversely proportionate to size, an animal will expend roughly similar amounts of energy (per gram of tissue) by the time it dies whether it is a mouse, a human being, or an elephant.

Despite its importance, size control (as the problem is known) remains one of the most poorly understood areas of biology, likely because of the vast number of variables that affect it. Even the genetic approach, which has proved so useful in discovering the genes that control most aspects of development, is of limited use in understanding size control. New tools and, more importantly, new concepts are needed if we are to understand how animals regulate biological proportion. Until that happens, size control will remain one of the many black boxes of nature.

EMBRYOS IN MOTION

MOST OF OUR UNDERSTANDING OF EMBRYOGENESIS COMES FROM studying snapshots; developmental biologists isolate embryos—whether wild-type or mutant—at various stages of development, which they then examine in great molecular detail. But a decade ago, a lecture by Caltech biophysicist Scott Fraser changed how I thought about morphogenesis by emphasizing one simple facet of development: the element of time.

Like many budding scientists and engineers, Fraser spent hours as a young person tinkering with electronics—dissecting audio equipment and other devices and then reassembling them in creative ways. As an undergraduate, he studied the physical properties of biological membranes until, as he describes it, he "made the mistake of looking through a microscope." The subject of this happy blunder was an embryo, and Fraser found himself enthralled by its beauty and resilience. In that moment, Fraser became a developmental biologist, using his command of gadgetry to design a new generation of microscopes.

It was not the technical virtuosity of Fraser's microscopes, impressive as they were, that stuck with me from his talk. Rather, it was the simple notion that looking at still images of embryos—as embryologists often do—gives an incomplete and possibly incorrect impression. To make his point, Fraser showed the audience a set of color photographs taken during a football game, pictures showing (1) helmeted players running about; (2) players heaped together as a human pile; (3) a single player with one leg hyperextended; (4) players wearing different jerseys bent at the waist and facing each other across an imaginary line. Fraser then pointed out the obvious: anyone without knowledge of football would be hard-pressed to understand the game's rules or even pick out the important events.

Next, he showed several short, poor-quality videos, each no more than 10 seconds long—videos showing a snap, a handoff, a tackle, and a punt. Now, there was context to the images, a sense of cause and effect. The added dimension of time had given a framework for understanding the images that the still pictures lacked.

"Even bad movies of bad games tell us a lot more about what goes on," Fraser said.

Of course, he was right—the easiest way to learn the rules of a game is to watch others play it (or to play it yourself). Yet we rarely study development that way, relying instead on lifeless material extracted from embryos at given stages. There are advantages to this more static approach, in that it permits a deep dive into developmental mechanisms. But something is lost when we look solely at snapshots.

Making high-resolution movies—a method called "live imaging"—can fill the gap, but it is fraught with challenges. First, the embryo cinematographer must capture the actions of each cell in detail (close-ups) without losing the broader view (wide-angle shots). Moreover, live imaging is expensive, as the hardware and other resources needed to film embryos in informative ways must be customized and optimized for each experiment. But perhaps the biggest impediment is simply getting a microscope objective where it needs to be—the "optically hostile" environments of the embryo, as Fraser calls them. Mammalian embryos are the hardest to film, given that most of development takes place within the protection of the uterine wall. But once these technical hurdles are resolved, embryo filmmaking stands to revolutionize our understanding of morphogenesis.

Another of the technique's pioneers is Nicolas Plachta, a colleague of mine at the University of Pennsylvania who studies early mammalian development. As we discussed previously, a mammalian embryo's first "decision" is determining which cells will become inner cell mass (future fetus) and which will become trophectoderm (future placenta). We know that a blastomere's position in the morula (inside vs. outside) and its level of Hippo signaling (high vs. low) play a role, but these are not the sole factors determining the outcome. By filming embryos at the 4-, 8-, and 16-cell stages, Plachta and his team found that the inheritance of certain cytoplasmic proteins—keratin-8 and keratin-18—was a strong predictor of a cell's later fate. Specifically, blastomeres that inherited these keratin proteins went on to become TE, while those that lacked the keratins became ICM. (Further experiments showed that these effects were not merely correlative, but that adding or removing these keratins from certain cells altered their destinies.) This is but one piece of evidence that the uneven distribution of certain proteins points cells in one direction or another. It is a conclusion that would have been hard to reach by means other than live imaging.

As John Sulston learned as he mapped the cellular lineage of *C. elegans* one cell division at a time, the simple act of observing can provide vast amounts of information. Remarkably, what took Sulston and colleagues months to accomplish can now be performed in a matter of days using new methods for image processing and sophisticated molecular tools. All of this bodes well for the future of embryo studies. For if a picture is worth a thousand words, then certainly a movie is worth a thousand pictures.

WE BEGAN THIS BOOK WITH THE ONE CELL PROBLEM—THE INCREDIble yet indisputable fact that every animal begins life as a single cell—and, in the past few chapters, we've gotten a taste of nature's solutions to the problem. We have seen how genes provide a framework for cellular differentiation while gene regulation leaves room for interpretation of these heritable directives. We have come to appreciate how the embryo uses cell-cell communication to balance the deterministic impulses of a cell with its adaptability. And we have just learned that cells acquire a sense of direction, allowing them to construct three-dimensional tissues as complex as an eye, a kidney, or a brain.

Of course, we have merely scratched the surface. While the number of molecular messages guiding development is not overwhelming—amounting to several hundred transcriptional regulators and signaling factors—what cells really care about is the additive effects of these instructions. Because each combination can have a distinct meaning, each cell must continuously work its way through trillions of calculations to find its way in the cellular society. As a field, developmental biology is thus undergoing a transition—shifting from a focus on single genes and single pathways to a consideration of gene networks and protein "interactomes," the ripple effects that spread out when a single gene or protein is altered. This more integrated way of looking at cells, tissues, and organisms is known as "systems biology," and it relies on massive computers and mathematical modeling to approximate the processing power of a single cell. So if you find yourself wondering how, after all the territory we have covered so far, all of this comes together—what almost-magical set of chemical reactions allows life to exist and persist—then you are walking the boundary of our current understanding, beyond which lie discoveries yet to be made.

Regardless of the mysteries that still remain, one thing is already clear: embryonic development provides an important window into human disease. Embryonic signals that stimulate the growth of fibroblasts, which give tissues their firmness, cause fibrosis in adults, the tissue scarring that precedes organ failure. The growth programs that expand one cell into the billions that make up a newborn baby can, if inappropriately activated, lead to cancer. The genes regulating programmed cell death contribute to disorders as varied as neurodegeneration, heart disease, and autoimmunity. This, in the end, is the blessing and curse of our biology—that the same processes nature uses to construct the body are all too often the means of its undoing.

Developmental biology is history, and as such, it requires no practical justification. But if one were to need a reason to study the embryo, the medical knowledge that has emerged from developmental studies should satisfy any such requirement. Despite its deficiencies, our understanding of developmental mechanism is now rich enough for it to be useful. The juxtaposition between the beauty of the basic and the utility of the applied is what makes learning from the embryo so gratifying, and it is this duality that we will explore next.

INTERMEZZO: MATURATION

The first thing that hits you when you enter a biomedical research building is the smell. In my building, a 14-story tower built in the 1990s, you notice it before you even cross the threshold—a peculiar combination of sterilized chow and bedding material drifting up from the vivarium in the basement, where mice and other animals are housed. But this aroma, a mix of hay and burnt pie, is only an introduction. Other scents await, ranging from the putrid to the sweet to the antiseptic.

First come the yeast labs, permeated by the fruity bouquet of a brewery. The source of the fragrance is *Saccharomyces cerevisiae*, brewer's yeast, a single-celled organism that (unlike bacteria) houses its DNA inside a nucleus. This subcellular attribute grants *S. cerevisiae* membership in the exclusive club of *eukaryotes* (Greek for "good nut"), a feature that makes yeast more closely related to humans than to *E. coli* and an ideal subject for studying the universal cellular properties of gene regulation, cell division, and metabolism.

Next, we come to the laboratories using cultured animal cells. Here, the trace scents of bleach and ethanol, intended to keep the environment sterile, are unmistakable. Much of the work takes place in designated "tissue culture" rooms—spaces removed from the main laboratory space so that bacteria, yeast, fungi, and other microorganisms are less likely to contaminate the more finicky animal cells.

At last, we come to the developmental biology labs, where the scenery varies with the types of embryos being studied. A *Drosophila* lab can be distinguished from a *Xenopus* lab or a mouse lab by the differing infrastructure, foods, and materials that sustain each species. And of course, the labs have different scents—the fly lab bears the odor of cornmeal, the frog lab smells like an aquarium, and the mouse lab, as we already experienced, has hints of rodent bedding and grub.

As we tour the building, we are in a sense like paleontologists working our way down an exposed rock formation, the fossils in each layer reflecting different chapters in the story of evolution. But instead of being arranged according to epoch, like the remains in a geological dig, the subjects of biomedical research are distributed more haphazardly. A lab using fish may find itself flanked by a lab employing mammalian cells. Researchers studying cell signaling, cancer, and neurodegeneration can coexist side by side. While some superficial differences (like smell) distinguish these enterprises, their similarities are far more numerous, for the biology of life comes down to the same fundamental chemistry. Consequently, biomedical researchers rely on a common set of tools regardless of the questions they are pursuing—freezers, centrifuges, water baths, incubators, microscopes, pipettes, and bags upon bags of plastic tubes.

There is another sense in which we have, up until this point, emulated paleontologists: our emphasis, like theirs, has been on history. Specifically, students of evolution scrutinize animal remains to determine how new species evolved from preexisting ones, while we, as students of the embryo, seek to understand how a fertilized egg "evolves" into a newborn animal. To be sure, the two disciplines employ different tools and concepts. Evolutionary biologists use fossilized skeletons to identify historical branchpoints; we use cellular pedigrees. Rather than viewing natural selection as the driving force behind new forms, we attribute that property to cell signaling. We also operate on vastly different timescales. But whether viewing an evolutionary trajectory or a developmental one, we all seek to learn from the past.

At this point, however, our path diverges from that of the fossil hunters. The most obvious difference between our enterprises, of course, is that as developmental biologists, we study living creatures. This is a great privilege, and it comes with an equally great responsibility—the imperative for those of us who use animals for research, especially vertebrates, to take that responsibility seriously, treating them humanely and acknowledging their sacrifices. But there is another and perhaps even more important difference between our approach and that of the paleontologists: our aspirations involve the future as well as the past. We are not bound by the historical records that constrain evolutionary biologists—petrified remains, DNA sequences, and geo-environmental details. Instead, we recognize that the embryo also holds lessons for times ahead, lessons that can reshape our individual and collective futures. We will now make it our business to realize its wisdom and apply it for good.

In a recent essay, biologist and historian of science Scott Gilbert argued that all branches of modern biology—from neuroscience to genetics—originated with the study of embryos. "Development was originally seen as the motor of evolution," Gilbert stresses, pointing out that in the late 1800s, the word "evolution" was as likely to refer to embryonic development as it was to the ancestry of species. The other foundational concept of the nineteenth century—cell theory—also has its roots in embryology, as the notion that all cells come from other cells germinated with Robert Remak's observations of cleavage divisions in frog embryos.

Other disciplines soon followed. Cellular immunology got its start when Russian zoologist Élie Metchnikoff, in his studies of starfish larvae, noted that cells eat other cells, a recognition of the critical immune process of *phagocytosis*. Theodor Boveri and Thomas Hunt Morgan, both dyed-in-the-wool embryologists, unleashed the field of molecular genetics with their recognition that developmental traits are encoded by chromosomes. Rudolf Virchow, the founder of modern pathology, observed that tumors arose "by the same law which regulated embryonic development," formalizing the close connection between cancers and embryos. And it was embryologist Ross Granville Harrison whose studies of early frog embryos established the central paradigm of neurobiology—the principle that neurons reach out across long distances to form connections (synapses) with target cells—inventing the technique of animal cell culture in the process.

Gilbert's depiction of embryology as a research wellspring is both apt and relevant for us, as everything we have learned about development so far has prepared us for the forward-looking material that awaits us in the upcoming chapters. The embryo regulates differentiation, gene expression, cell-cell signaling, and morphogenesis using tools that nature devised to surmount the One Cell Problem. Those tools, and others, are now being repurposed in the laboratory and the clinic, providing scientists and physicians with the opportunity to emulate and modify nature's handiwork in ways that never existed before. It is a power to preserve and create that we are just beginning to master.

Chapter 7

STEM CELLS

Mistakes . . . are the portals of discovery.

— JAMES JOYCE, *ULYSSES*

Ernest McCulloch and James Till could not have been more different. "Bun" McCulloch, as friends and colleagues called him ever since childhood, was stocky and studious, a child of privilege who had attended Toronto's best private schools. He paid little attention to his appearance, apart from his signature bow tie, and his clothes were often covered with the residual chalk dust from blackboard work he had done earlier in the day. His mind was a stew, constantly churning with new ideas, more interested in big concepts than practicality. Till, by contrast, was raised on a farm in rural Saskatchewan. He was an outdoorsman by nature, more athlete than intellectual, but at work he was meticulously dressed and punctilious. Where McCulloch was the academic, happiest when engaged in a discussion of poetry or history, Till was the pragmatist, the guy who could get things done.

To anyone who didn't know them, it would have come as a surprise that the pair would find any common ground, much less forge a lifelong collaboration. But as it happened, their differences would become their advantage. Where McCulloch thought in grandiose terms, focused on the forest, Till was a detail man, his eyes never losing sight of the trees. If McCulloch meandered too far out on an intellectual limb, Till was the one who would inevitably pull him back to safety. McCulloch knew that he could come up with as many wild ideas as he wished since Till would be there, serving as a check. At the lab bench, where it counted the most, the two complemented each other perfectly.

It was the force of the atom, of all things, that brought the two men together in the late 1950s. The twin holocausts of Hiroshima and Nagasaki that marked the end of World War II had demonstrated the horrors that splitting uranium and plutonium nuclei could precipitate, and the Cold War between the United States and the Soviet Union was fueling fears that more nuclear conflagrations could follow. Till and McCulloch were new to research—McCulloch had practiced medicine for nearly a decade before making the leap to the lab, while Till was fresh out of graduate school, having just earned a PhD in physics. But despite their lack of experience, their backgrounds made them ideal candidates for a new Canadian research unit devoted to finding treatments for radiation poisoning. It was there that McCulloch the dreamer and Till the go-getter would identify a new brand of cell—one that, like the zygote, could differentiate along many paths. A cell whose role in development, tissue function, and cancer has been a subject of aspirational science ever since.

ROUGHLY 500 MILLION YEARS AGO, GIVE OR TAKE 50 MILLION YEARS, nature began a grand experiment. Until that point, life had flourished mainly in the form of single-celled organisms—cyanobacteria, archaebacteria, algae, fungi, and other creatures that lived their lives as active but autonomous entities. The unicellular world was resilient, and organisms could be found everywhere, from the intense heat of a hydrothermal vent to the frigid cold of the tundra, where they subsisted on scarce organic chemicals and sunlight. The only attribute they shared was independence, each living a contented existence on its own.

Then, in a period of unprecedented diversification known as the "Cambrian explosion," nature challenged the status quo. It was a trial, really—a test to determine whether more complex organisms, creatures composed of many cells instead of one, would fare as well as their primitive ancestors. It wasn't the first time these biological limits had been tested. Nature had tinkered with multicellularity before, but these had been false starts, and the few creatures that had emerged left no phylogenetic successors behind. The Cambrian endeavor, nature's fresh attempt at multicellularity, could have failed just as easily. But for some reason, it didn't, and the millions of extant and extinct animal species that followed were the result.

Early multicellular animals were small, earthworm-like creatures that slithered from place to place, eating their way through the lawns of bacteria that covered the ocean floor. With time, more complex organisms arose—echinoderms, primitive snails, trilobites, and crustaceans, whose descendants spread out across the oceans. Multicellularity allowed organisms to gather their cells into collaborative clusters—primitive organs that let these creatures seek out food sources, recognize danger, assimilate nutrients, and remove waste. Most importantly, they could move farther and faster than a unicellular organism could, allowing the offspring of these new animals to crawl out of the sea some hundred million years later. Evolution was painting with its fullest palette.

The fact that multicellularity took so long to emerge—billions of years after the planet was sufficiently cool to support life—implies that huge hurdles impeded the formation of many-celled organisms. The issue was not one of cell proliferation, for nature had solved that problem—the mechanics of making a cell divide—well before the Cambrian explosion. Instead, the difficulty was cellular diversity. Uniformity (at the cell level) was acceptable in the unicellular world, where the most pressing goal was simply to make copies of the original. But in a multicellular world, variation became a priority. Heterogeneous cell types were needed to support idiosyncratic and complex jobs—absorption, secretion, sensation, and movement at first, followed by vision, hearing, and communication later—such that uniformity would no longer suffice. The leap to multicellularity demanded coordination between division and differentiation, an increase in both cell number *and* cell diversity. In short, evolution faced the One Cell Problem, the same conundrum we began with. A special type of cell with the ability to do both—the *stem cell*—was to become the centerpiece of nature's solution.

THE COBALT BOMB

THE YEAR WAS 1958, A TIME WHEN THE RELATIONSHIP BETWEEN SCIence and society was unusually complex. For many, there was a sense that technology would be able to crack humankind's oldest problems and shed light on its biggest mysteries. Previously fatal diseases could now be cured with new antibiotics or had become entirely preventable with life-saving vaccines. Advances in science and technology had revolutionized the farming industry, providing a means of feeding the world's exploding population. Scientists working on the smallest scales imaginable were defining the very substance of matter, while astrophysicists studying the stars were detecting cosmic echoes of the universe's origin. Science was a fountainhead, a seemingly inexhaustible source of new knowledge that would inevitably lead to better health and enhanced quality of life.

At the same time, technological advances had created an atmosphere of risk and dread. Just over a decade had passed since the end of World War II, a conflict that left no doubt concerning technology's ability to destroy. The atomic attacks on Japan had killed over 150,000 people—half instantaneously and half from the delayed effects of the blast and radiation. The subsequent Cold War, with its race to build bigger and more versatile instruments of destruction, had brought the planet to the threshold of annihilation more than once. If science held the key to solving society's ills, it might just as easily be the vehicle for its extermination.

Nowhere was this ambiguity more apparent than in the field of radiation research, where in addition to testing the atom's power to harm, scientists were also studying its ability to heal. A German physicist named Wilhelm Röntgen was the first to recognize radiation, reporting the existence of X-rays in 1895. His breakthrough was followed over the next few years by Henri Becquerel's discovery of uranium's radioactive properties and Pierre and Marie Curie's discovery of radium and polonium. By the turn of the century, physicians had begun experimenting with the invisible rays released by these materials as treatments. Their patients came with a variety of ailments, ranging from lupus to tuberculosis to cancer. Radiation was found to improve some cancer patients' odds of survival, but it quickly became apparent that radiation could also *cause* cancer. Early therapies were clumsy, causing as many side effects as cures, and the field stagnated.

That changed after World War II, when a 35-year-old Canadian physicist named Harold Johns began testing a new material—a radioactive metal called cobalt-60. This man-made form of cobalt, an isotope with radioactive properties, proved to be more versatile, reliable, and cheaper than radium, which had previously been the dominant source of radiation. After some calibration, Johns and his team treated their first patient in 1951—a mother of four with cervical cancer. The result was the complete remission of what had been an apparently incurable tumor. The success sealed Johns's reputation. In 1958 he obtained permission to begin treating patients with cobalt-60 at the Ontario Cancer Institute (OCI), a newly launched Canadian center dedicated to cancer research. The "cobalt bomb," as it came to be known, became a powerful symbol of the peaceful application of nuclear technology.

Sitting atop Toronto's Princess Margaret Hospital, the OCI had a mission that stood apart from some of the centers of scientific excellence that preceded it, places like the Stazione Zoologica or the Institut Pasteur, where we found ourselves earlier. Those institutes had a mandate that was purely academic—the pursuit of fundamental science of the highest quality. The OCI, by contrast, had a dual mission. Beyond simply advancing scientific knowledge for its own sake, OCI scientists were charged with doing research that would also have practical applications. Such "applied research," deemed less pure or high-minded by some scientists, didn't bother Johns, who saw no contradiction in asking scientific questions that might be relevant to human health. His own work in physics, the most fundamental of all the sciences, was a glowing example (figuratively, if not literally) of how basic research could benefit humanity.

It was against this Cold War backdrop of hopes and fears that Johns began building a physics division within the OCI, where his chief objective was to develop better radiation-based therapies for cancer. But equally important was the goal of reducing radiation's side effects, which not only would make such therapies more tolerable but would also save lives in the event of a nuclear conflict. Advancing these goals, Johns believed, meant populating the unit with scientists from diverse backgrounds—clinicians, biologists, and physicists—and forcing them to work together. For a siloed scientific world unaccustomed to such mixing, it was an unusual approach, but it was responsible for the unlikely pairing of McCulloch and Till, two of Johns's earliest recruits.

MCCULLOCH HAD BEEN IN CLINICAL PRACTICE FOR NEARLY A DECADE when he started working at the OCI. Medicine had been the family trade. McCulloch's father and two uncles were all doctors, and he had followed in their path, completing medical school at the University of Toronto and then settling into a career as a general practitioner with a specialty in hematology—disorders of the blood. The work kept him busy but unsatisfied. At heart, McCulloch was an intellectual, a scholar as knowledgeable of Cartesian philosophy as he was of diabetic neuropathy and as familiar with Roman history as with pulmonary tuberculosis. As a doctor in training, he had developed an interest in research during a medical-school laboratory rotation, and by the late 1950s, that interest in research had matured into a calling. When the opportunity to join the OCI came along, McCulloch leapt at it. Science satisfied his hunger for big ideas, for he saw it as a game of wits with nature in which imagination was the only constraint.

James Till came to the OCI by a very different path. He was five years younger than McCulloch, and his farm upbringing in Saskatchewan—

which sometimes had him working from five in the morning until nine at night—had made him strong and focused. He excelled at school, where his analytical acumen was obvious to his instructors, and after college he was admitted to Yale's competitive biophysics PhD program. Till could have landed any number of jobs in the United States after Yale, but his northern roots ran deep, and Harold Johns found it relatively easy to lure him back to Canada. Even at the height of professional activity, when he and McCulloch were unraveling the secrets of stem cells, James Till continued to return home every fall to help with the harvest.

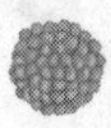

DESPITE THE EARLY SUCCESSES OF THE COBALT BOMB, THE TREATment was still toxic. Moreover, no one was quite sure how it worked. Conventional wisdom held that cancer cells are more sensitive to radiation than normal cells, and clinical observations agreed with this belief. The cobalt bomb, when it was successful, appeared to dissolve tumors with relatively less collateral damage to normal tissue. But in 1956, a Colorado biologist named Ted Puck challenged this notion by exposing either normal cells or cancer cells to different doses of radiation. Puck reported that a given dose killed normal cells and cancer cells with similar efficiency, a result which indicated that normal cells may be just as susceptible to radiation as cancer cells. The claim met with skepticism: If cancer cells and normal cells were equally sensitive, why didn't radiation therapy cause more damage?

The question intrigued McCulloch, who suspected the answer lay in the fact that Puck's studies were performed using cells grown in incubators—a method of in vitro (Latin for "in glass") culture that might give a false impression of a cell's sensitivity. By contrast, all the evidence that radiation was more effective against cancer cells had come from patients, where the exposure occurred within the body, or in vivo (Latin for "in a living thing"). No one had directly measured the radiation sensitivity of cells inside a living animal before, and McCulloch saw this as an ideal way to get his feet wet in the laboratory. Such a project, merging basic science with clinical applications, was precisely the kind of thing Johns had in mind for his team.

To ensure cross-disciplinary collaboration, Johns had instituted a rule for his unit: any biologist doing experiments with radiation had to be teamed up with a physicist. (Physicists, Johns believed, were far more quantitative than biologists, and forcing such pairings would ensure that whatever data were collected would be of the highest quality.) So Till became McCulloch's physicist, and McCulloch became Till's biologist. The two hit it off, neither one knowing that their partnership would end up going in a direction having little to do with the original question.

FINDING AN ASSAY

BIOLOGY IS A SCIENCE OF *ASSAYS*—THE EXPERIMENTAL SYSTEMS USED to collect data. Measuring an organ's weight is an assay, as is determining the amount of protein in a sample, the growth or death rates of cells, or the sequence of nucleotides that make up a gene. Assays may be low tech or high tech, requiring a simple scale or a sophisticated microscope. But they all have one thing in common: an ability to make reproducible measurements. Robustness is an assay's most important feature, and thus the assay must be tailored to the inquiry at hand. For Till and McCulloch, who wished to measure how radiation affects a cell's viability inside the body, this meant developing an assay of their own.

For an initial grasp on the impact of radiation on the body, McCulloch didn't need to look any further than the Japanese victims of the atomic bomb. More than a third of the casualties from "Little Boy" and "Fat Man"—the two nuclear devices detonated over Hiroshima and Nagasaki—had survived the initial blasts, only to die two or three months later from the horrific syndrome of tissue degeneration known as radiation poisoning. Of the organ systems involved in the disease, the blood was most severely affected, for once a victim developed severe anemia—a deficiency of oxygen-carrying red blood cells—death was all but certain. Consequently, radiation biologists, fearful that the decade-old Cold War with the Soviet Union could turn hot at any time, focused their attention there, hoping to find ways to mitigate the blood-poisoning effects of radiation damage. Understanding how radiation crippled the blood system was not merely a question of academic interest—it was a matter of national security.

During World War II, a Chicago-based physician named Leon Jacobson had been tapped to be the chief physician for the Manhattan Project, the secret race to split the atom for use in warfare. Jacobson's day job was to monitor the health of the staff who worked with radioactive materials, but in his off hours, he maintained his own research program, using animals to perform the kinds of studies that could never have been conducted on people. When Jacobson began his research, it was already known that all blood development—or *hematopoiesis*—occurred within the bone marrow, the meaty pulp in the cores of long bones. But as his work proceeded, Jacobson began to wonder whether the bone marrow was the only site capable of generating new blood cells. Perhaps some other organ might pick up the slack, compensating if the bone marrow were too sick to do so. Moreover, he had a strong candidate for this "backup" system—a nonessential abdominal organ called the spleen.

Jacobson's first step was to define the "lethal dose" of radiation—the dose that would irreversibly, and fatally, obliterate the blood system. That dose, he discovered through trial and error, is between four and nine "gray" (the standard unit of absorbed radiation) in most animals he tested. But when he exposed such "lethally irradiated" mice to the same radiation dose while simultaneously protecting their spleens with a lead shield, the animals survived. Something about the spleen was safeguarding the blood system against the deadly effects of radiation.

Jacobson's findings caught the attention of Egon Lorenz, a German scientist who had fled to the United States before the war. Jacobson believed his findings could be explained by a "humoral factor" (what today we might call a hormone)—a putative molecule, made by the spleen but carried to the bone marrow, where it could act to rescue blood development. Lorenz had different ideas. He believed that Jacobson's results indicated that the spleen itself could harbor blood-producing cells. The bone marrow might be the major source of hematopoiesis, Lorenz reasoned, but it need not be the *only* source. Perhaps other sites, including the spleen, could serve as factories for blood production in times of stress.

This hypothesis, in turn, sparked another question—if blood-producing cells could exist in either the bone marrow or the spleen, could they also move freely about the body? The ultimate test of such an idea would be to show that bone marrow cells, transferred from one mouse to another, could give rise to a new blood system, and Lorenz did precisely that—crippling the bone marrow of one mouse (the recipient) with radiation, then injecting a slurry of bone marrow cells from another mouse (the donor). Lorenz's conjecture proved correct: mice that received the transplanted cells were rescued from certain death.

The technique—*bone marrow transplantation*—revealed an important fact about the blood system: the cells that give rise to the blood are mobile, retaining their ability to differentiate and function even after being extracted, manipulated, and reimplanted. This property could even extend across species, as the bone marrow from a rat could, in some circumstances, rescue a lethally irradiated mouse. Bone marrow transplantation would eventually become a state-of-the-art treatment for leukemia and other blood disorders, but for Till and McCulloch, the technique would serve a more modest purpose. It was the in vivo assay they needed to determine the blood-harming effects of radiation.

LUMPY SPLEENS

BUN MCCULLOCH COULDN'T SLEEP. SOMETHING HE HAD SEEN IN THE lab that day was weighing on him, and he was still trying to make sense of it. Ten days earlier, he and Till had performed a typical bone marrow injection, which involved infusing tens of thousands of cells into the bloodstream of irradiated mice. Their goal was to derive a "survival curve"—a graph that could describe with precision how many cells were needed to rescue a lethally irradiated mouse. With such a standard in hand, the two scientists could start testing interventions, pretreating the transplanted cells, or the recipients, with various chemicals to see which, if any, might improve the chances of survival after exposure to radiation. But such trials were a long way off. For now, the bone marrow injections were merely controls, designed to refine the in vivo assay. The protocol was relatively simple: irradiate the host, inject various quantities of cells, and wait. If, a month later, the mouse still lived, this meant that the number of cells injected had been sufficient to rescue the bone marrow. If the mouse had died, the number had been insufficient. In all cases, the spleen would be examined as part of the protocol. The experiments were routine, time consuming, and, after so many replicates, somewhat boring.

On the day in question—a cold Toronto Sunday in 1960—McCulloch was doing his standard laboratory rounds. He and Till ran a small operation, and they took turns coming in on the weekends to keep an eye on the mice and prepare for the next week's experiments. It should have been a short day for McCulloch, as the animals weren't due to be sacrificed and examined for another 20 days. The researchers knew from their previous experience that it took a full month to determine whether a transplant was successful or not. But on this particular Sunday, for whatever reason, McCullough deviated from procedure—he decided he would examine the animals early, nearly three weeks before the experiment's planned termination date.

In his two years at OCI, McCulloch hadn't ceased to marvel at the anatomical kinship of mouse and human—the striking conservation of organ position and function. While a mouse's heart may beat 5 to 10 times as fast as a human being's, and it may breathe 10 times as often, these differences melt away at the cellular level. The mechanisms used by the heart's contractile cells (myocardium) to pump blood are nearly identical in the two species, and the unimaginably thin membranes that line the alveolar spaces of the lung—separating the harsh external world from the body's precious interior—have the same molecular makeup in rodent and primate. Whatever lessons McCulloch might extract from these experimental creatures, there was no doubt in his mind that those lessons would translate to people.

These very thoughts might have been going through McCulloch's mind as he gave each mouse a fatal overdose of anesthetic and proceeded with his autopsies. Suddenly, something caught his eye. The spleen, normally a smooth tongue-shaped organ with a bright reddish hue, was studded with bumps. He dissected a second mouse, and then a third. All of them had the same oval gray nodules, some with 1 or 2 and others with 8 or 10. They were so big and out of place that at first McCulloch wondered if they might be metastatic tumors, clusters of disseminated cancer cells that had landed there from some other part of the body. But *metastasis* to the spleen was unusual, and these mice did not have cancer, at least not as far as McCulloch knew. Something else was going on.

How easy it would have been for him to dismiss these bumps as an irrelevant oddity. The readout of the experiment, after all, was simply whether a mouse lived or died, a measurement that could later be used to measure radiation sensitivity. The lumpy spleens, therefore, had nothing to do with the question he and Till had set out to answer. Had he simply waited the full 30 days after transplantation, as he was supposed to, the spleens would have looked normal. Which left him with the most likely explanation—the bumps were simply a bizarre epiphenomenon, an irrelevant distraction from the important work he and Till were doing. If he had any sense, he would put the result out of his mind and return to the comfort of the cozy fire waiting for him at home.

FORTUNATELY, BUN MCCULLOCH'S MIND DIDN'T WORK THAT WAY. HE could no more ignore an unexpected result than an English teacher could overlook a sloppy conjugation. Still, it makes one wonder how many would-be discoveries have been shelved before seeing the light of day, victims of a scientist's overzealous focus. I myself am guilty of this sin of omission—times when I have set aside an inconvenient result, only to find later that what I had attributed to flawed technique instead reflected something original or important. It is true that on balance, the surprise finding—an outcome that falls far afield of expectations—is more likely to reflect a mistake than a breakthrough. But, rarely, such unintended circumstances can be a spyhole into the unknown. As science-fiction writer Isaac Asimov reflected, "The most exciting phrase to hear in science, the one that heralds new discoveries, is not 'Eureka!' but 'That's funny . . .' " He was right, but it works only if the observer is smart enough or brave enough to follow up.

Instead of ignoring the lumpy spleens, McCulloch doubled down. He completed his autopsies, carefully recording the number of nodules in each mouse. Then, he opened the laboratory notebooks in which Till had recorded the details of his injections and compared the two lists—one holding the number of cells Till had infused and the other recording the number of splenic lumps he had just counted. There was an unmistakable pattern—the two numbers were almost perfectly correlated. The more bone marrow cells Till had infused, the more lumps McCulloch observed. He didn't know what this meant, but the trend was too strong to be the result of chance.

The next morning, McCulloch walked into the OCI waving his graphs at Till. He confessed to terminating the experiment early and described his observations and conclusions. The physicist had to agree—the correlation was intriguing and no coincidence. Now, Till's physics background came into play. He recalled Jacobson's wartime experiments—how the Manhattan Project physician had rescued mice from the toxicity of radiation by shielding their spleens. Could the spleen, that oblong organ of unknown purpose, have been acting as a factory yet again, each lump representing a separate site of production for new blood cells?

There was an important difference between Till and McCulloch's experimental setup and the one Jacobson had used. Because the Toronto mice had been exposed to lethal doses of radiation that completely destroyed their bone marrow, they lacked the capacity to regenerate the blood system on their own. Therefore, the new blood system had to have been derived from the donor marrow. In a standard experiment, which lasted a month, the infused cells journeyed to the recipient's bones, where they kick-started hematopoiesis anew. But by interrupting the recovery process early, could McCulloch have discovered an intermediate cellular rest stop? Had the infused cells paused in the spleen long enough to begin making new blood, a process that might have manifested itself as a series of lumps? This possibility, if true, could explain the phenomenon's timing. The nodules appeared a week and a half after irradiation, when the need for new blood was at its greatest, but they were gone three weeks later, when the bone marrow had resumed its role as the primary site of blood production.

During his dissections, McCulloch had set aside several nodules, placing them in a formaldehyde fixative to keep all the tissue's cells in place. Now, he and Till began preparations to investigate the lumps—slicing them with a precision cutting device (a microtome) to make the paper-thin sections that would reveal the lumps' cellular composition. The whole procedure took less than a day, and the tissues, when viewed under the microscope, gave an answer as obvious as the tie adorning McCulloch's collar. The nodules (which they renamed *colonies*) were indeed factories of blood production, small bone marrow–like aggregates bursting with activity.

McCulloch had chanced upon a new biological phenomenon—one he hadn't known existed until a day earlier—and it cried out for explanation. Had each colony arisen from many cells or from one? The latter possibility was far and away the more exciting prospect, as it would mean that a single cell could give rise to all the components of the blood system. This was a radical idea, as the zygote was thought to be the only cell known to be capable of generating diverse cell types. And yet McCulloch could feel in his bones—down to the marrow—that each colony had arisen from one cell.

A SPECIAL KIND OF CELL

WITH AN AVERAGE LIFE SPAN OF ALMOST 80 YEARS, HUMANS FALL IN the middle of the animal longevity spectrum, a range bordered on one end by worms and flies that live for days to weeks, and on the other end by whales and clams that can live for more than a century. For small animals with a short life expectancy, nature didn't have to plan for long-term care, since the organs of these animals could be counted on to function long enough to spawn the next generation. But for larger animals, those that live for years or decades, evolution had an additional consideration: cellular turnover.

Most cells have a limited life span, whether due to the wear and tear of daily life or programmed cell death, and long-lived organisms require a fail-safe way to replenish their depleted ranks. Rates of cellular turnover can vary widely among tissues—from the brain, where cells persist for decades, to the intestine, where most cells last only a week or two. It is in these tissues with rapid turnover that stem cells provide a never-ending supply of fresh recruits.

The term "stem cell" was coined in the late nineteenth century by German zoologist Ernst Haeckel. Haeckel was Hans Driesch's thesis adviser, and he is most famous for asserting that "ontogeny recapitulates phylogeny"—the exaggerated claim that embryos pass through successive stages of evolution during development, as if built layer by layer before arriving at a final product. When it came to stem cells, or "*stammzellen*," as he called them, Haeckel had a similarly larger-than-life idea; to him, a stem cell was the founder of all animals and all tissues—a primordial seed from which the entire tree of life had sprouted.

In the early twentieth century, another German naturalist—Ernst Neumann—reinvented the term. Neumann had a simpler definition: a "stem cell" was any cell that could give rise to many other types of cells. Neumann was among the first to realize that all the cell types composing the blood—oxygen-carrying red blood cells, infection-fighting white blood cells, and clot-forming platelets—were produced in the bone marrow. Indeed, years of staring at slides of bone marrow had given Neumann a sense of how these three different lineages developed. In 1912, he speculated that all were descended from a common cellular ancestor, what he called "the great lymphocytic stem cell."

Neumann had no way to prove his theory, as this would require a technology he lacked—a way to prospectively isolate such cells and demonstrate their developmental potential. Moreover, Neumann's model clashed with the conventional wisdom, which held that red blood cells, white blood cells, and platelets each arose independently. The great lymphocytic stem cell fell into disfavor, remaining untested until Bun McCulloch resurrected it nearly five decades later.

IF TILL AND MCCULLOCH'S INSTINCTS WERE CORRECT—IF A SINGLE stem cell was indeed giving rise to each of the spleen colonies—such cells were incredibly rare. McCulloch's graphs showed that an infusion of 30,000 cells yielded an average of three colonies per spleen. This meant that the presumptive stem cells constituted roughly 1/100th of 1 percent of all the cells in the bone marrow; finding them would be nearly impossible unless one knew precisely what to look for. At the same time, such cells would have to be remarkably prolific, for each colony contained tens if not hundreds of thousands of cells.

But before such speculations could be entertained, the first order of business was determining whether each colony was indeed derived from a single founder cell. It was a question that bore a striking resemblance to the quandary that John Gurdon was facing around the same time across the Atlantic Ocean—proving that a single transplanted nucleus could generate a whole new frog. Gurdon's approach had been to use a genetic marker, one that could discriminate between two species of frog, allowing him to distinguish donor nucleus from host egg. Till and McCulloch would have to find a different cell-labeling technique. Ultimately, their approach relied on the same method they had employed all along: exposure to radiation.

Although invisible, radioactive particles pack a mean punch. Each particle carries a substantial amount of energy, ripping through anything that stands in its way. After a radioactive particle and DNA collide, the cell does its best to repair the damage, but it is easily overwhelmed. Consequently, radiation exposure leaves a genetic scar—an identifiable change to one or more of a cell's chromosomes. Once permanently fixed in the genome, these abnormal configurations pass from parental cell to daughter and granddaughter, creating a record of the cell's history of damage. But, critically for Till and McCulloch, no two cells exposed to radiation will have the same pattern of damage—the collisions (and the cell's attempts at repair) are simply too random.

For Andy Becker, a graduate student in Till's lab, these unique radiation scars held the key to understanding the origins of the spleen colonies. His plan involved irradiating the bone marrow of a donor animal, using a quantity of radiation much lower than the lethal dose needed to kill off the recipient's bone marrow. This would mark each of the donor cells with a distinctive chromosomal pattern—a pattern that would persist if and when they repopulated the host. The logic was simple. If every cell in a colony carried the same chromosomal irregularities, then they must have come from a single cell—evidence that the colony was *clonally* derived. But if cells of a colony carried different radiation scars, then the colony must have been derived from multiple founders—evidence that it was *polyclonally* derived.

Becker had some legwork to do before he could get started. Most importantly, he had to find a quantity of radiation that was high enough to induce detectable patterns of chromosomal damage but not so great as to kill the cells or prevent their engraftment. Through trial and error, Becker discovered the sweet spot—a dose approximately two-thirds of the one Till and McCulloch had used to lethally irradiate a mouse. At this level of exposure, distinctive patterns of chromosomal breaks emerged in roughly 10 percent of the cells, enough to serve as unique identifiers, or fingerprints, of cellular lineage.

After some additional fine-tuning, Till, McCulloch, and Becker geared up to repeat the marrow-transplant procedure; only this time, they would be transplanting marked cells. As they hoped, the spleens of these transplant recipients were littered with colonies a week and a half later, their cells bearing distinctive patterns—an easily scored chromosomal excess or deficiency. While each individual colony bore a different chromosomal signature, the cells *within* a colony carried the same chromosomal scar. The fingerprinting method had worked—each regenerating cluster, bursting with red blood cells, white blood cells, and platelets, had arisen from a single cell.

Till and McCulloch's first papers—the ones in which they had measured the radiation sensitivity of normal bone marrow cells and reported their observations of the lumpy spleens—had been published in *Radiation Research*, a respected but obscure journal, where they went largely unnoticed. But Becker's work, published in *Nature*, warranted the highest billing. Despite its clunky title—"Cytological Demonstration of the Clonal Nature of Spleen Colonies Derived from Transplanted Mouse Marrow Cells"—the paper was a birth announcement of sorts, notice of the stem cell field's arrival.

OUT OF ONE, MANY

IN ADDITION TO PROVIDING THE FIRST DEFINITIVE EVIDENCE THAT stem cells exist, these experiments also laid out their essential characteristics. The modern definition of a stem cell has come to include two key features: *multipotency* (a cell's ability to give rise to multiple different kinds of cells) and "self-renewal" (a cell's ability to make more of itself). These powers—to multiply and to diversify—are enabled by a special property of stem cells known as "asymmetric cell division."

In practical terms, asymmetric cell division provides a means for increasing the number and diversity of cells at the same time. When a stem cell divides, it gives rise to two daughters, just like any dividing cell. But what sets the stem cell apart is its ability to create daughter cells that are unalike. When a stem cell divides asymmetrically, it gives rise to a daughter that is a carbon copy of itself and another daughter that is completely different, all in a single division. The ability of a stem cell to determine its legacy in this manner gives it a nearly unlimited capacity to spawn new progeny with varying characteristics. In this sense, a stem cell is the cellular equivalent of a fairy-tale character who, having been granted three wishes for rescuing a genie from its bottle, uses one to wish for more wishes.

The blood turned out to be a first-rate example of a stem cell–derived tissue. The lineages of the blood—red blood cells, white blood cells, and platelets—carry out vastly different jobs, and each can trace its ancestry back to a series of precursors known as *progenitor cells*. For example, red blood cells are derived from progenitors called "erythroblasts," while platelets are derived from progenitors called "megakaryocytes." But if one traces these progenitors back even further, one ends up at the source—the *hematopoietic stem cell*—resulting in a cellular "hierarchy." Each time a stem cell divides, it provides the raw material needed to produce new blood while replenishing its own stem cell reservoir.

Stem cells beget stem cells.

"You have to realize what exponential growth is like," McCulloch later said, describing cellular dynamics in terms that would make a gambler salivate. "If you had one dollar and you doubled it twenty times, do you know how much you'd have? A million dollars!"

The Toronto scientists had proved the existence of stem cells, but isolating them was another matter entirely. Despite their remarkable properties, stem cells had no distinguishing features they knew of, nothing that would cause them to stand out and be recognized. "Going from one cell to a million cells in twenty doublings, it would be next to impossible to find your original dollar—or cell—among the million," McCulloch said. It was their behavior—not their appearance—that made the cells so special.

LIKE SO MANY ADVANCES IN BIOLOGY, THE DISCOVERY OF STEM CELLS was a matter of serendipity. An impatient scientist terminated an experiment earlier than he was supposed to, sending himself and his colleagues in search of an explanation. Till and McCulloch happened to be working on the blood, but would the same serendipity eventually have struck other scientists working in other organs?

Perhaps not. There is little doubt that stem cells are responsible for renewing tissues in which cells are turning over constantly; these include the blood, the intestine, and the skin. But the role of stem cells in other tissues—especially those with lower rates of cell turnover—is more variable. In some body parts, like skeletal muscle, stem cells do little under normal circumstances and are called into action only when muscles are injured. In other organs, like the pancreas, liver, and kidney, there is scant evidence that stem cells exist under any circumstances; in these organs, new cells simply come from the division of preexisting cells. And within a given tissue, stem cells reside in *niches*, specialized tissue sanctuaries that reinforce the stem cell's special properties—serving up a kind of stem cell energy drink. Whereas blood stem cells have an easy time finding their niche, stem cells in solid organs (that is, tissues other than "liquid" blood) have no way to do so, and these must be studied by means other than transplantation. It was therefore perhaps inevitable that the blood stem cell would become the first cell, other than the zygote, shown to exhibit multipotency and self-renewal.

Till, McCulloch, and Becker had no way to purify these cells—a feat that even now, some 60 years later, requires great effort. But this didn't hold the field back. In some sense, there was even something satisfying or at least symmetrical about the researchers' inability to lay hands on their experimental subjects; for years, high-energy physicists had dealt with the same constraint, forced to study radioactive particles without ever seeing them. Researchers around the world, many of them the scientific progeny of Till and McCulloch, could still study the properties of these hematopoietic stem cells (or HSCs, as blood stem cells are more accurately referred to) without purifying them.

The inability to purify the cells didn't matter from a practical standpoint either. In the late 1960s, E. Donnall Thomas, a Cooperstown, New York–based physician inspired by these stem cell pioneers, began working out the conditions needed to transfer bone marrow stem cells from one human being to another. There are significant differences between Till and McCulloch's technique and the clinical protocols used in contemporary bone marrow transplantation—now referred to as "hematopoietic stem cell transplantation" (HSCT)—which required the incremental optimization that is the bread and butter of clinical research. But with the innovations those clinical researchers provided, HSCT became a standard medical procedure. To date, more than 1.5 million transplants have been performed, and the number—currently 90,000 HSCTs per year worldwide—continues to climb.

In the late 1980s and 1990s, another type of stem cell would come to occupy the spotlight, a cell far older in evolutionary terms than the HSC and possessing a far greater potential. This was the "embryonic" stem cell, with its breathtaking and nearly limitless ability to both multiply and specialize. These attributes made it the secret ingredient in nature's recipe for multicellular life—the solution to the One Cell Problem that made animal life on the planet possible. And those same ancient properties—the stem cell's ability to give rise to any cell type in the body—now stand to transform the future of medicine.

Chapter 8

CELLULAR ALCHEMY

The universe is full of magical things patiently waiting for our wits to grow sharper.

— EDEN PHILLPOTTS, *A SHADOW PASSES*

In 1990, an Italian-born developmental biologist named Mario Capecchi made history by creating the world's first "knockout mice." The term may evoke images of rodents with prodigious boxing skills, but the recipient of the decisive blow in a knockout mouse is not an opposing mouse; it is a gene. Through a painstaking process of cellular engineering, Capecchi had worked out the conditions to alter the mammalian genome, and now he was telling the world how he had done it.

Capecchi had an unimaginably difficult childhood. Born in Italy to a single mother just before World War II, he became homeless at the age of four when the Nazis arrested his mother for anti-Fascist activities. The toddler was shuttled from one living situation to another, first to a farm, then an orphanage, and finally into the care of his estranged and abusive father. After the war, Capecchi's mother found him in a Reggio Emilia hospital, close to death from typhoid and malnutrition. From there, mother and son moved to the suburbs of Philadelphia, where his uncle, a physicist and Quaker, took them in. The traumas of wartime dissipated slowly, as Capecchi immersed himself in sports and schoolwork. He attended Antioch College, a small liberal arts school in Ohio, spending summers in MIT biology labs through the college's work-study program. From there, he attended graduate school at Harvard (under Jim Watson, of double-helix fame), and ultimately landed a job at the University of Utah.

This was the early 1970s, and recombinant DNA technology was advancing at full speed. In his new lab, Capecchi sought to improve the methodology of *gene transfer*—getting foreign DNA to insert itself permanently into a cell's genome. Existing methods were inefficient, reliant on a cell's innate tendency to take up pieces of DNA. But Lawrence Okun, one of Capecchi's new colleagues in Salt Lake City, had a suggestion for his new colleague: Why not inject DNA directly into the cell's nucleus?

Capecchi decided to give it a try. He figured out how to make needles with a diameter of 1/10,000th of a millimeter or less, the fine dimension necessary to deliver the foreign nucleic acids into the cell's inner sanctum. To his delight, Capecchi saw that Okun had been right; he could inject DNA more efficiently than a cell could ingest it, and within a year he was up to nearly a thousand injections per hour. At that rate, he and his team could investigate how, and where, foreign pieces of DNA integrated into the host cell's genome. For the most part, the integrations were random—the injected DNA fragments seemed to insert themselves indiscriminately into one or more chromosomes. But every so often, a DNA fragment managed to find its way to its corresponding location in the genome—the singular chromosomal position where the sequence of the injected DNA and the host chromosome were the same (assuming, of course, that the injected fragment contained such a corresponding sequence).

This phenomenon, called *homologous recombination*, is a cell's natural editing mechanism, how it remedies the inevitable mistakes that occur each time it copies the billions of nucleotides in its genome. During homologous recombination, cells swap out a faulty piece of DNA by using the copy on its other chromosome—a correct one—as a template, or master. This proofreading exercise is part of a cell's routine, occurring with every round of division. But Capecchi's experiments suggested that this gene-swapping behavior was not limited to a cell's native DNA. A cell could be tricked into thinking a foreign nucleic acid, introduced from the outside, belonged in its genome, inserting it precisely where it should be.

Homologous recombination became a way to substitute the normal copy of a gene with one designed in the laboratory—a molecular imposter replacing the natural version. If the engineered piece of DNA didn't differ too much from the native gene—a few nucleotides here or there—the cell would claim it as its own. To verify that the method was working as intended, Capecchi performed the procedure with easy-to-score markers, genes whose products, when interrupted, made cells sensitive to an antibiotic. Using such tricks to refine the procedure and calculate its efficiency, he and his students estimated that homologous recombination occurred in roughly 1 out of 1,000 injected cells.

Capecchi was not alone in these gene-swapping efforts. Oliver Smithies, a British-born scientist who had set up shop at the University of Wisconsin, was pursuing a similar line of work. Smithies and his colleagues had used a comparable protocol to create cells with mutations in the beta-globin gene—a gene whose product forms part of the oxygen-carrying protein hemoglobin. Simultaneously competitors and allies, Capecchi and Smithies shared a vision: both believed that the next step should involve disrupting genes in living animals, not just in cultured cells. If such an effort was successful, future biologists wouldn't have to rely on nature to deliver mutants. Instead, scientists could decide for themselves which genes they meant to disrupt and then mutate them at will.

Homologous recombination alone would not accomplish this goal. Capecchi and Smithies could create designer mutations in cultured cells, but engineering such errors in an animal—where they would be carried by all the creature's cells—was a taller order. Further progress was needed, an innovation that would allow genetic alterations to make their way into the germline—an animal's heritable genome. By chance, such a technology was developing in parallel, with its origins in cancer.

A CURIOUS TUMOR

IN THE MID-1950S, ROY STEVENS, A GENETICIST WORKING AT THE Jackson Laboratories in Bar Harbor, Maine, noticed unusual tumors growing in the testes of mice. For years, the laboratory's scientists had bred the rodents to create "inbred strains"—stocks of mice akin to the purebred pea plants Mendel used to work out the principles of genetics. Because members of an inbred strain are genetically identical, sharing characteristics like coat color, they are especially useful for immunological studies. Cells transplanted between members of the same inbred strain are accepted without difficulty, whereas those grafted across strains are rejected as foreign.

The tumors Stevens observed appeared in an inbred strain called "129," and they were *teratomas*, a rare but fascinating type of tumor. Unlike most malignancies, whose cells have a more consistent microscopic appearance, teratomas are a cellular Noah's ark, encompassing the gamut of cell types normally present in the body. Muscles and nerves, cartilage and bone, epithelia, fat, even hairs and teeth—all these tissue types can be observed in a single teratoma. This breadth of cellular differentiation raised the prospect that these unusual tumors might share a kinship with normal embryonic development. Other observations supported this idea too. Stevens's tumors had originated in the testes, the source of sperm, and human teratomas arise most often in the testes, or in the ovaries, the source of eggs. Perhaps these facts were not coincidental.

Stevens found that if he ground up the cells from one tumor and injected the slurry into the flanks of mice of the same strain, new teratomas grew, proof that the tumors were transplantable. He then began to vary his transplantation technique, leading to an experiment in 1968 whose startling result further solidified the connection between teratomas and development. Instead of tumor cells, the donors in this experiment were normal mouse zygotes or two-celled embryos, which he then implanted into the testis of a recipient mouse. Remarkably, the transplanted embryos grew into teratomas. This implied that the testis was fertile ground for the development of this sort of tumor, even from a normal cell.

Two years later, Croatian biologist Ivan Damjanov and colleagues took Stevens's experiment one step further by taking more advanced mouse embryos, those that had progressed to the epiblast stage, and transplanting them into a completely different location—the kidney. Again, teratomas grew. Embryos that would have matured into newborn mice had they been left alone to develop in the uterus—their cells ordinary in every respect—formed tumors when displaced. The line between embryogenesis and carcinogenesis was obviously fuzzier than it appeared; being in the wrong place at the wrong time had dire consequences.

THIS LATENT CARCINOGENIC POTENTIAL PRESENT IN OTHERWISE normal embryonic cells reinforced the idea that teratomas recall some long-lost developmental program. Stevens came to view his tumors almost as mini-embryos, masses comprising two cellular compartments: (1) a differentiated population, consisting of muscle, bone, hair, and other specialized cell types, and (2) an undifferentiated population, lacking specialized features, from which the first population descended. Indeed, it was hard to imagine how the tumor could achieve such a diversity of cells in any other way. He concluded that the tumors must employ a developmental hierarchy—a succession of differentiated cells arising from undifferentiated cells. Most tantalizing of all was the possibility that each tumor had arisen from one cell.

This idea appealed to Barry Pierce, an American pathologist at the University of Michigan who had taken a deep interest in Stevens's work. To test the notion that each teratoma had arisen from a single multipotent cell, Pierce and his student Lewis Kleinsmith began implanting individual teratoma cells into recipient mice. Out of 1,700 single-cell transplants, approximately 40 grew into tumors. It was a herculean effort, with a low rate of success. But it was proof that an individual teratoma cell could spawn the vast diversity of cellular citizens within these tumors.

Researchers were now referring to Stevens's cells as *embryonal carcinoma cells*, or ECs, and parallels with other research areas began to appear. John Gurdon had recently learned that cloning a frog (using nuclear transplantation) grew harder as the nuclear donor became more differentiated, aligning with Stevens's and Pierce's idea that undifferentiated teratoma cells retained greater potential than their differentiated counterparts. Moreover, ECs bore the hallmarks of the hematopoietic stem cells Till and McCulloch had discovered in their bone marrow transplantation experiments: the properties of multipotency (the capacity to spawn many types of differentiated cells) and self-renewal (the capacity to make more of themselves). But there was an important difference. Unlike Till and McCulloch's stem cells, whose potential was limited to the blood, cells from Stevens's tumors could become almost anything.

EMBRYONIC STEM CELLS

STEVENS DISTRIBUTED HIS TUMOR-BEARING 129 MICE WIDELY, SHARing them with researchers who wished to investigate the tumors' unique properties on their own. One of the beneficiaries of Stevens's generosity was Ralph Brinster, a veterinarian and developmental biologist at the University of Pennsylvania who had pioneered much of the methodology for culturing and manipulating mammalian eggs and embryos. Brinster had shown it was possible to create "hybrid" embryos by taking cells from one embryonic blastocyst and injecting them into the blastocyst cavity of another embryo. To distinguish donor-derived and host-derived cells, Brinster had used strains with different coat colors, making the amalgamated offspring—called *chimeric mice* after the part-lion, part-goat creatures from Greek mythology—easy to spot, their mottled coats revealing their dual parentage.

One of the outstanding questions about ECs was whether there were limits to their developmental potential. If it was truly as vast as Stevens believed, then their capacity to give rise to diverse cells should extend beyond tumors, to normal development. Brinster's hybrid embryo system could provide an answer to the question: Would Stevens's EC cells, injected into normal host blastocysts, incorporate into a hybrid embryo and differentiate as if they were normal cells?

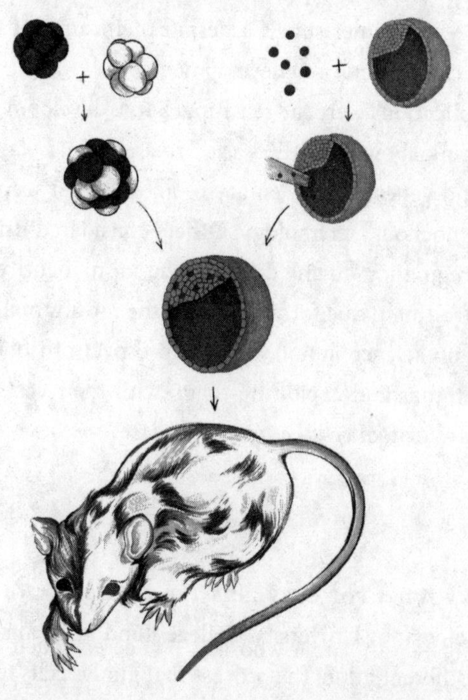

Chimeric mice can be created in multiple ways. On the left, two morulae, derived from two strains of mice with different coat colors, are allowed to fuse and grow to the blastocyst stage. On the right, ECs or other stem cells are microinjected into a blastocyst cavity (again using strains of mice with distinguishable coat colors), where they incorporate into the inner cell mass. The resulting hybrid blastocysts are then implanted into the uterus of a recipient mother, where they can develop to term. The resulting offspring will consist of descendants from both sets of starting materials–in effect, making them the products of three or four parents!

To answer this question, Brinster injected teratoma cells from strain 129 mice, with their gray-brown hair, into blastocysts isolated from white (albino) mice. Again, the process was inefficient. But Brinster succeeded in recovering one chimera—an animal with patches of dark hair decorating its otherwise white coat. It was a signal that the embryonic environment could quell the tumor-forming instincts of the EC cells, the inverse of Stevens's earlier experiments with normal embryonic cells (which turned into cancers when relocated to the wrong environment). Other researchers soon replicated Brinster's findings, achieving even greater degrees of chimerism. ECs, despite their cancerous origins, could indeed participate in normal development.

The implications were far-reaching. Some developmental biologists started to think about using ECs as a Trojan horse, a way to deliver foreign cells and genes into an embryo—an idea that would later evolve into "gene knockout" technology. Others considered using the cells to study gene regulation during development, or to identify and study the signals controlling tissue formation. But the most tantalizing notion of all, verging on science fiction, was the prospect of isolating ECs from human teratomas and exploiting them, with their vast developmental potential, as a form of *cell therapy* for disease.

ANOTHER REQUEST FOR STEVENS'S TERATOMAS CAME FROM MARTIN Evans, a biologist at University College London. Evans was interested in cellular differentiation, the process that guides cells to their places in the cellular society. He was using frog embryos for his studies, but he had grown discouraged with the complexity of this experimental system. Too much was happening too quickly in the embryo—most of it obscured from view and at a minuscule scale that was impractical for biochemical studies—to make sense of anything. Evans reasoned that an in vitro approach, one in which cellular differentiation could be studied in a tissue-culture dish, would be better, and the EC system seemed made to order.

In 1969, Evans wrote to Stevens, who shipped a cage of tumor-bearing 129 mice to England straightaway. The first step was generating "cell lines" from the tumors—immortal lineages that could be expanded indefinitely in culture (rather than transplanting the cells from mouse to mouse, as Stevens had been doing). Compared to normal cells, cancer cells grow like proverbial weeds, the very feature that makes them a medical scourge. But this creates a research opportunity, as tumor cells—once chopped up, plated in petri dishes, and left to grow in an incubator—adapt to culture far more readily than their normal counterparts. With Gail Martin, a newly recruited postdoctoral fellow, Evans established several immortal stocks of ECs, using "feeder cells," a second type of cell included in the culture, to help sustain them.

Soon, Evans and Martin were testing the differentiation properties of the EC lines. By every measure, the cells behaved as if they had come straight out of a fresh tumor. Cells injected into the flanks of mice yielded teratomas, while those injected into the blastocyst cavity gave rise to chimeric mice. Even single cells could give rise to teratomas, reproducing Pierce and Kleinsmith's earlier findings and indicating that the EC lines retained their developmental potential. But there was a critical difference—unlike the tumors from which they had been derived, these EC cells could be maintained indefinitely in incubators or as frozen reserves, called into service whenever needed.

Despite contributing to embryonic development in the form of chimeric mice, there was one barrier that the ECs could not cross: the germline. Evans's most fervent hope had been that within the chimeras, some of the animals' sperm or egg cells would be derived from the injected ECs. If this were the case, the EC genome could be passed along to the next generation through those gametes. But this never happened. Out of the hundreds of offspring of chimeric mice Evans examined, none bore the telltale brownish hairs of strain 129. The EC cells could contribute to many cell types in the chimeras, but the gametes, it appeared, were off limits. It was an ironic letdown. EC cells had originated in the testes, the site of male fertility, yet they were sterile themselves.

Evans, who by now had moved to Cambridge University, concluded that if he wished to derive stem cells that could pass their genes to the next generation, he would have to go to the only source that could be counted on to produce offspring—the embryo itself. In 1980, after working with EC cells for a decade, Evans turned his attention to the mouse blastocyst, betting that these cells would do what the cancer cells could not.

After months of trial and error, Evans succeeded in getting blastocyst-derived cells to grow in culture. Like the EC cells, these noncancerous cells formed multiple lineages in culture and gave rise to tumors when injected into mice. Importantly, they lacked the kinds of chromosomal abnormalities that, it turned out, had made it impossible for the EC cells to pass along their genes. Within months, Gail Martin, Evans's former postdoc, accomplished the same feat in her own laboratory in San Francisco. Martin referred to the new cell lines as *embryonic stem cells*, or ESCs, and the name stuck.

Technically speaking, ESCs do not exist in nature. Although the cells from which they are obtained—cells residing in the blastocyst in vivo—have similar properties, they normally differentiate within a day or two as development proceeds. In vitro, by contrast, ESCs can remain in a state of suspended (undifferentiated) animation for weeks, primed to execute one of a thousand different specialized programs when released from the culture conditions that prevent their differentiation.

Within a couple of years, Evans, Martin, and others demonstrated that these new embryo-derived cells could do everything the ECs could do, and more. Most importantly, chimeric mice made from them could pass their genomes along to the next generation. It was a property that elevated their status from being multipotent (capable of giving rise to many cells in the body) to *pluripotent* (capable of giving rise to all cells in the body). The era of embryonic stem cells had finally arrived.

I HAD MY FIRST GLIMPSE OF ESCS IN 1991, EARLY IN MY PHD. BY THEN, ESC methodology had made its way around the world, and Michael Shen, a postdoctoral fellow in the laboratory where I worked, had mastered these protocols and was using them to study differentiation. A brilliant but introspective scientist, Michael's demeanor was usually quite serious. But one morning when I came into the lab, he had an almost mischievous smile on his face.

"Want to see something really creepy?" he asked, pointing to a microscope on whose stage he had placed a circular tissue-culture dish.

In culture, cells find their footing on the plastic surface of such dishes, deriving their nourishment from the nutrient-rich medium placed on top. Cultured cells live, grow, and differentiate in specialized incubators whose conditions (especially temperature and acidity) are designed to mimic those inside the body. Although they look nothing like their in vivo counterparts, cultured cells can still be classified based on their appearance, magnified, in the dish. Fibroblasts, connective tissue cells that respond to cuts, like to spread out across the surface of a tissue-culture plate, as if attempting to seal a perceived breach in the plastic. By contrast, epithelial cells, whose role in the body is to create an impermeable barrier, form a flat layer of closely connected cells, giving the appearance of a cobblestone-lined street.

Peering through the eyepieces, I saw some cell types that I recognized and some that I didn't. I could easily make out the fibroblasts, spread across the dish like a patterned tablecloth. (These were fibroblast "feeders," nurse-like cells added to the culture to keep the ESCs from differentiating.) In addition, there was another cell type. Aggregates of tightly packed cells appeared to have settled on top of the fibroblasts, forming cellular mesas rising from the fibroblast plains. As my eyes grew accustomed to the view, I focused on one of the clusters, when suddenly, to my amazement, I saw movement—kinetic waves that pulsed from one end of the structure to the other every few seconds. The colony was *beating*.

Fibroblasts are supposed to keep ESCs from differentiating, but clearly these feeders had done their job imperfectly, for the ESCs within some of the colonies—including the one I was watching—had assumed the specialized identity of cardiac muscle cells, or cardiomyocytes. They beat because of the internal pacemaker all heart muscle cells possess. While this pulsing made their differentiated state easy to recognize, more sophisticated ways of detecting differentiated cells would have revealed many other lineages in the dish as well—neurons, cartilage, intestine, and more.

"Pretty cool, huh?" Michael said.

Two thoughts came to mind as I watched this cluster of cells go through round after round of contraction. First, I simply marveled at what I was seeing—cells trying to behave as if they were a part of a heart. Contractions that began on one side of the colony propagated to the other, a consequence of electrical signals conducted from one cardiac cell to the next. These cells had no way of knowing that they were out of place—that they belonged inside a heart, not a dish—but they didn't seem to care. At the same time, I thought about the origin of the colony, which had started out as a single ESC that happened to land in that singular spot I was fixated on. It seemed implausible, almost impossible, that its descendants could have made such a long developmental journey—from embryo to heart—in a dish. And yet here they were, beating, their rhythmic contractions an unambiguous testament to a successful passage.

KNOCKOUT MICE

THE ABILITY TO GROW EMBRYONIC STEM CELLS WAS THE TECHNICAL breakthrough Capecchi was waiting for. In 1985, just before Christmas, he traveled to Cambridge, England, where Evans taught him how to derive and culture ESCs and make chimeric mice. All the elements were then in place to create strains of mice with altered genomes: Capecchi would engineer mutations in the ESCs, these cells would be used to make chimeric mice, and the chimeric mice would pass the mutations along to future generations.

Which gene should he target first? Capecchi thought deeply about the question, as success or failure could hinge on picking the right one. The front-runner was the gene encoding an enzyme known as HPRT. Mutations in HPRT cause Lesch-Nyhan syndrome, a human disease that affects mostly males and leads to gout, developmental delay, and, often, self-mutilating behavior. For Capecchi's purposes, HPRT had several valuable attributes. The enzyme's activity could be readily detected, making it easier to demonstrate whether the gene had been "knocked out" or not. Furthermore, the HPRT gene is located on the X chromosome (explaining why Lesch-Nyhan syndrome affects mainly males)—meaning that, if he used male ESCs, with their single X chromosome, Capecchi would need to target only one copy for the cells to become HPRT-deficient.

Capecchi and his trainees—postdocs and students—began putting the pieces together. They engineered fragments of DNA with faulty HPRT genes and fed them to ESC cells, which took up the flawed fragments and inserted them into their own genomes. As expected, most cells introduced the foreign DNA randomly, adding the nucleic acids haphazardly to the billions of bases already present in the nucleus. But with a measurable frequency, the DNA integrated at the position of the normal HPRT gene, replacing the functional copy with the faulty one. In parallel, Oliver Smithies had shown that it was possible to use homologous recombination—the gene-swapping technique—to correct a naturally occurring mutant version of the HPRT gene. Homologous recombination could create designer mutations, but it could also fix them.

Smithies, continuing to work with the HPRT gene, succeeded in getting his engineered ESCs to "go germline," the final step in making a knockout mouse. The research teams began targeting other genes as well, and in 1990, Capecchi's team took the final step—using the mutant ESCs to make mutant mice. The gene they chose to knock out this time was called *int-1*, and its disruption resulted in mice with a variety of neurological phenotypes.

These advances turned the field of genetics on its head. For decades, geneticists had gone from mutant to gene, identifying flies, worms, or mice with interesting phenotypes and then working their way back to the DNA culprit. This was the "genetic approach," the method that the Heidelberg scientists used to identify pattern-shaping genes and that Brenner, Sulston, and Horvitz used in their quest to understand cellular behaviors. Now, for the first time ever, researchers didn't need to rely on nature to produce mutants; they could choose for themselves which gene they wished to alter. Homologous recombination and ESC technology made it possible to go from gene to mutant—to elucidate the function of a gene by altering its sequence and asking what phenotypes come of it. The strategy was called, quite aptly, *reverse genetics*, and it spread across the field of biology like wildfire. By the time Evans, Capecchi, and Smithies were awarded their Nobel Prizes for gene knockout technology in 2007, more than 10,000 mouse genes had been disabled by homologous recombination—genes whose biological functions touch on every disease afflicting humankind.

FROM MOUSE TO HUMAN

GENE KNOCKOUT TECHNOLOGY CONTINUED TO IMPROVE, AND WITH further modifications it became possible to control the timing of gene loss. Beyond the standard knockout, which obliterates a gene (and hence its protein product), researchers began making subtler mutations—changes affecting a single nucleotide in a gene's DNA sequence that more closely resemble changes associated with human disease. Virtually every gene in the mouse genome has now been mutated in some way, with many altered several times over.

In the early 1990s, researchers began to think more seriously about the therapeutic potential of stem cells. If mouse ESCs could form any cell type given the right signals (which occur naturally in living animals), then certainly their human counterparts should be able to do the same thing, creating a nearly unlimited potential to treat human disease with engineered cells. That, above all else, became the field's ambitious new goal—to make human ESCs that could replace cells lost to heart attacks, strokes, kidney failure, emphysema, autoimmunity, neurodegeneration, burns, metabolic diseases, and numerous other conditions.

Researchers presumed that generating human pluripotent cells would be relatively straightforward, believing that Evans and Martin had already done the heavy lifting. But translating stem cell protocols from mouse to human turned out to be more complicated. Human embryos were, for some reason, less cooperative, and it became obvious that the derivation of human ESCs would require new approaches. After several years, two researchers—Jamie Thomson at the University of Wisconsin and John Gearhart at Johns Hopkins—resolved the technical issues, and by the late 1990s, there was a renewed sense that human ESCs would transform the practice of medicine, with their potential to fix a broken heart, or kidney, or spinal cord.

But the stem cell field faced another hurdle: getting the cells to behave. Before human ESCs could usher in a new era of cell-based therapies, researchers would have to derive the various types of cells needed to treat disease. This proved to be quite difficult. Differentiation, as we saw earlier, is the outcome of multiple cellular conversations—the convergence of myriad signals whose exact combinations, provided at the right time at the right level, guide cells to their proper place in the cellular society. To achieve the same end point with ESCs, it would be necessary to reproduce some or all of those signals.

Hoping to generate such usable cells, stem cell researchers began exposing ESCs to various protein and chemical cocktails—some recipes devised empirically, and others based on knowledge of embryology. Progress came gradually. Scientists refined their differentiation protocols, filling their dishes with cells resembling those normally present in the pancreas, heart, lung, eye, kidney, liver, intestine, cartilage, bone marrow, brain, and other tissues.

But the field has yet to fully reach its intended goal. Most stem cell researchers will confess that even today, the products of ESC-differentiation protocols lack the functionality of their normal cellular equivalents in the body. We don't understand why such cells behave inferiorly to their naturally arising counterparts, but the difference suggests that certain factors present in the body are missing from the tissue-culture dish. Likewise, we still don't understand why an ESC fulfills its normal destiny when injected into a blastocyst, and yet that same cell forms a tumor when injected into the flank of an animal. These questions are a reminder of how little we still know about self-renewal, differentiation, and the One Cell Problem.

THE WAR AGAINST STEM CELLS

BY THE YEAR 2000, AN EVEN BIGGER STUMBLING BLOCK EMERGED—religious and ethical objections stemming from the fact that human ESCs come from human embryos. While there was not much outcry over using mouse embryos to make stem cells, destroying a human embryo was a different story. The very idea was unacceptable to millions of people, especially those who were unconditionally opposed to abortion. To them, it was irrelevant that the embryos—tiny balls of a hundred cells or so—were donated by couples who had undergone in vitro fertilization and no longer needed them.

On August 9, 2001, President George W. Bush spoke to the nation about human stem cell research. It was just a month before the attacks of September 11, and the speech tackled what *Time* magazine called "the central predicament of his young presidency." Only three years had passed since Thomson and Gearhart had announced the successful creation of human ESCs, but this was long enough for the field to find itself in the crosshairs of the American abortion debate.

Broadcast from his ranch in Crawford, Texas, Bush's address framed embryonic stem cell research as an issue that "lives at a difficult moral intersection, juxtaposing the need to protect life in all its phases with the prospect of saving and improving life." As he thought about the issue, Bush explained, he "kept returning to two fundamental questions. First, are these frozen embryos human life and therefore something precious to be protected? And second, if they're going to be destroyed anyway, shouldn't they be used for a greater good, for research that has the potential to save and improve other lives?"

The specific question Bush struggled with was whether tax dollars should be used to fund research that used the products of destroyed embryos, even if those embryos were "leftovers" from fertility clinics. His resulting policy decision pleased neither proponents (scientists and patient groups) nor opponents (abortion foes) of human ESC work. Bush's compromise was to allow the National Institutes of Health (NIH) to continue to fund research utilizing any of the 60 or so human ESC lines that had been created prior to his evening speech, but to ban the use of federal dollars for research on human ESC lines created subsequently.

Opponents of human ESC research called the Bush policy "morally unacceptable" for allowing any work involving human embryos to proceed. But for those who had pinned their hopes on the potential of stem cell research, Bush's compromise was devastating. The policy, with its allowances for a few dozen cell lines, made it sound like human ESC work could continue without significant disruption. But this was not the reality. Many of the existing lines were unsuitable for research—either they failed to grow in culture or didn't behave like stem cells. Other lines were proprietary, unavailable to the broader research community. And the few lines that remained did a terrible job of representing the diverse genetic backgrounds of humanity.

Paradoxically, the Bush policy may have accelerated the pace of stem cell research. In the months following Bush's speech, several states responded by sponsoring their own independent stem cell research initiatives. Private foundations and patient-advocacy groups, incensed that the government would close the spigot and deprive this promising field of vital dollars, also chipped in, supporting the creation of stem cell research institutes in the United States and abroad. It is hard to quantify the amount of funding generated through such initiatives. Nevertheless, it is quite possible that in the aggregate, the amount of money raised from nonfederal sources has exceeded what the NIH would have spent on human ESC research had there been no restriction on federal spending.

These political and ethical battles continued for the next two years. But in 2006, the landscape changed in ways that affected both the controversial aspects of the research as well as the science. For it was then that a Japanese surgeon and his student discovered another way to make pluripotent human cells—a method that didn't touch an embryo.

THE PHILOSOPHER'S STONES

SHINYA YAMANAKA WAS BORN IN OSAKA IN 1962, THE SON OF AN ENGINEER. Like many of the developmental biologists we have met previously—Roux, Miescher, McCulloch, and others—Yamanaka was a doctor before he was a scientist. He attended medical school in Kobe, specializing in orthopedic surgery. But along the way he concluded that he could do more to ease suffering as a researcher studying the causes of disease than he could as a doctor treating those who were already sick. He enrolled in a pharmacology PhD program, earned his degree in 1993, and then moved his family to San Francisco for a postdoctoral fellowship at the Gladstone Institutes, where he learned how to make knockout mice.

Upon returning to Japan, Yamanaka came to realize that the philosophy of science in the United States was quite different from what it was in his home country. In the States, the research environment had fostered creativity, risk taking, and vigorous intellectual interactions. In Japan, the metrics were far more practical. His colleagues had little interest in basic research, and they recommended that he switch to a project more closely tied to medicine. Finding it difficult to publish papers or get funding, Yamanaka flirted with abandoning his research aspirations and returning to clinical medicine.

Fortunately for him, and for medicine, the surgeon-scientist secured a new job at the Nara Institute of Science and Technology in Japan, where he was hired to run a mouse knockout "core facility." There, his job would be to make mice with designer mutations, altering the genome in whatever ways his colleague-clients desired. (Core facilities leverage economies of scale, providing services that are too expensive or too technically demanding for individual research labs to perform on their own.) He was good at generating knockout mice, and this made him valuable. His scientific self-confidence grew. At last, he had room to explore—the freedom and resources to pursue a project of his own.

By then, it was 1999, the year after Thomson and Gearhart defined the methods for making human ESCs. Yamanaka worked with mouse ESCs daily in his "knockout core," and he shared the widely held belief that human ESCs would one day revolutionize medicine. Stem cells, he decided, would be the subject of his independent research project. But unlike other researchers in the growing field, who were seeking out better ways to get stem cells to differentiate, Yamanaka would try to accomplish the reverse—turning a differentiated cell into a stem cell.

OF COURSE, OTHERS HAD ALREADY ACHIEVED A SIMILAR FEAT. JOHN Gurdon had been first, showing (via nuclear transplantation) that the developmental clock could be rewound all the way back. Gurdon's achievement had enabled the cloning of mammals, beginning with Dolly the sheep, and it showed that every cell holds the potential to become any other kind of cell. Then, just as Yamanaka was settling into his position at the Nara Institute, fellow Japanese researcher Takashi Tada showed that pluripotency could be induced by fusing a mouse ESC with a fully differentiated lymphocyte. This was critical, for it was a sign that the pluripotent state was "dominant." An ESC, with its far-reaching potential, could *reprogram* another cell, one already set in its ways.

Yamanaka knew that approaches like nuclear transplantation would never replace ESCs. Not only was the method technically demanding, but it also destroyed embryos in the process, the very impediment that had haunted the field all along. He imagined that there must be an easier way to turn back the developmental clock, a method for reprogramming a cell's nucleus that didn't require such highly specialized techniques. Scanning the literature for more straightforward alternatives, he settled on transcription factors—the gene regulators Jacob and Monod first revealed with their studies of bacteria and phage. With their ability to turn genes ON and OFF, the right combination of these DNA-binding proteins just might be able to reprogram a differentiated cell to pluripotency.

On its face, the idea seemed simply impossible. Dozens if not hundreds of transcription factors dictate which combination of genes is expressed in each cell, thereby determining its identity. The notion that he or anyone else could precisely substitute the right combination of transcription factors to convert a differentiated cell into a stem cell appeared hopelessly optimistic. It was an undertaking that, on first principles, shouldn't work.

Still, Yamanaka had his reasons for thinking it might not be so complex after all. For one, the dramatic changes in cell identity seen in flies with homeotic transformations—deranged *Drosophila* with an extra set of wings, or legs in place of their antennae—could be caused by mutations in a single transcription factor. Moreover, Harold Weintraub, a researcher at the Fred Hutchinson Cancer Center in Seattle, had shown that MyoD—a transcription factor involved in muscle differentiation—could convert a fibroblast into a myoblast (a muscle cell precursor). If a single transcription factor could redirect differentiation in such dramatic ways, perhaps a small number of transcription factors, in the right combination, could induce *de-differentiation*, forcing a specialized cell to stem-ness.

The first step was to assemble a list of all transcription factors that might be capable of doing the job. Scanning the literature, and relying on information gleaned from knockout studies, Yamanaka and his lab members compiled a list of 24 transcription factors that were either expressed specifically in ESCs or important for their function. If his supposition was correct, some or all of the genes on the list should be able to nudge a cell toward a stem cell–like state.

Then, it was time to test the hypothesis. Yamanaka, who by that time had moved his laboratory to Kyoto University, convinced graduate student Kazutoshi Takahashi to take on this risky project. Takahashi began by isolating each of the genes encoding these transcription factors and introducing them, one at a time, into mouse fibroblasts, cells that were specialized for wound healing. The hope was that he would see some change in the fibroblasts' appearance, an indication that they had developmentally reverted.

The results were uniformly negative—there was nothing to indicate the cells had taken any steps toward pluripotency. It was a minor disappointment, but not all that surprising. Not ready to give up, Yamanaka and Takahashi then wondered whether they might have greater success by combining the transcription factors. Takahashi thus introduced all 24 genes into the fibroblasts simultaneously, and this time, the result was different and truly remarkable—some of the cells resembled ESCs!

Some combination of transcription factors had rewound the developmental clock—but which ones? Were all 24 genes required, or could a subset get the job done? Through trial and error, Takahashi narrowed down the list to only 4 genes. When these 4 genes—now commonly referred to as the "Yamanaka factors"—were expressed in fibroblasts, the cells not only looked like ESCs, they behaved like ESCs: (1) they expressed the full range of ESC genes; (2) they made teratomas when injected into the flanks of mice; and (3) most important of all, their descendants incorporated into every embryonic tissue following their injection into mouse blastocysts.

The manipulated cells seemed equivalent to embryonic stem cells in all respects except one—they did not come from embryos. To emphasize this fact, and to skirt the controversies that had dogged the human ESC field, the researchers called this cellular rejuvenation process "induced pluripotency," referring to the products of their protocol as *induced pluripotent stem cells* (iPSCs).

Within a year, Yamanaka's lab used these same genes to create human iPSCs, a feat duplicated by ESC pioneer Thomson using a slightly different collection of genes. More than a dozen labs soon followed, showing that it was not just fibroblasts that could be reprogrammed to pluripotency; the developmental clock was reversible in a wide variety of differentiated mouse and human cells. The fact that so many independent (and skeptical) labs were able to replicate the results in such a short amount of time showed just how easy it was to restore a cell to its embryonic ground state.

Initially, some scientists worried that iPSCs might lack some of the potential of ESCs or differ in some other way, making them unsuitable for research or eventual clinical use. But concerns that iPSCs are a poor substitute for ESCs have dissipated over time, as it has proved difficult to find significant differences between iPSCs and ESCs. While ESCs continue to be used for some studies, the pendulum has shifted in favor of iPSCs; they are relatively easy to generate and can be derived from living cells taken from anyone. And because they bypass the ethical issues that overshadowed the ESC field in the first few years of the 21st century, they face few political or social obstacles. An experiment that never should have worked made it possible to create a one-of-a-kind embodiment of every human being, budding with potential.

CELLULAR AVATARS

IPSCS HOLD GREAT PROMISE. BUT THE FIELD IS STILL YOUNG, AND THE precise role these stem cells will play in the future of medicine remains undefined, a topic we will take up again in the final chapters. Nevertheless, pluripotent cells have already established their utility in "disease modeling"—a way to study illness without the risks and costs of human clinical research.

Animals, especially mice, have been the traditional subjects of disease modeling. But the results have not always been satisfactory. Although mice and humans diverged only 75 million years ago, a blink in evolutionary terms, disease often magnifies their differences. Because of variations in cellular composition, metabolic pathways, and molecular structures, drugs that work in mouse models often fall short in human patients. Similarly, mice engineered to carry a mutation that causes severe disease in humans may, with surprising frequency, suffer no apparent ill effects. The challenges are even greater with infectious diseases, as viruses are extremely specific when it comes to which organism(s) they choose to infect. For all they have taught us about embryogenesis and physiology, mice can be hit or miss when it comes to understanding illness or developing new therapies.

Pluripotent cells have emerged as an alternative, and sometimes more reliable, model of disease. Methods for making iPSCs were relatively straightforward from the start, and the protocols have been further simplified over the past decade. Currently, a researcher can generate iPSCs—starting from a blood or skin sample, or even from cells in urine—in a matter of weeks. Thousands of iPSC lines have been derived, each an avatar of the individual from whom it was generated. They run the gamut of racial and ethnic backgrounds and disease states, some held privately in research labs and others in "iPSC banks," where they can be shared among researchers through institutionally sanctioned procedures.

Many disease-modeling efforts have focused on neuromuscular diseases—conditions like Alzheimer's disease, Parkinson's disease, muscular dystrophies, and others—where better treatments are desperately needed. Animal models of these diseases have been disappointing, and patient samples are hard to come by, as these diseases often affect the brain. Hence, scientists have turned to iPSCs as an alternative way of studying human illness. For example, Harvard researchers Clifford Woolf and Kevin Eggan have used iPSCs to study a rare inherited form of amyotrophic lateral sclerosis (ALS, or Lou Gehrig's disease) in which motor neurons no longer function properly. Using approaches like Yamanaka's, the researchers created iPSCs from patients with this rare form of ALS and then, using a cocktail of factors, induced them to differentiate into cells resembling motor neurons. The patient-derived motor neurons exhibited abnormal electrical behaviors that were absent from motor neurons derived from control iPSCs, indicating that these cultured cells did indeed reflect features of the human condition. The researchers then looked for drugs that could correct the electrical defect and found one—an anti-epilepsy drug—that did so. This, in turn, has led to a clinical trial.

The liver has also received attention from iPSC disease modelers. The liver has many functions, from regulating metabolism to synthesizing blood proteins, and these activities are carried out by hepatocytes, the dominant type of cell in the liver. But hepatocytes—even those that have been freshly obtained from a biopsy—are difficult to grow in culture. Over several decades, researchers worked out methods for differentiating iPSCs into hepatocyte-like cells, opening up several research opportunities. For example, iPSC-derived hepatocytes could be used to predict whether a molecule is likely to damage or kill hepatocytes—a common drug side effect known as hepatotoxicity.

iPSC technology can also be used to repurpose old drugs. An example of this application comes from liver researcher Stephen Duncan, who used iPSCs from patients with familial hypercholesterolemia (FH)—an inherited disease leading to massively elevated bloodstream levels of LDL cholesterol—to identify new ways to treat high cholesterol. (Among its major functions, the liver controls levels of LDL cholesterol, the so-called bad cholesterol that can promote cardiovascular disease and death when elevated.) Duncan found that when he differentiated iPSCs from FH patients into hepatocytes, they exhibited the same abnormal behavior they would have in the clinic, including the inappropriate release of large quantities of the LDL-related protein apoB. Duncan's team then asked whether any existing drugs might be able to inhibit the cells' production of apoB. The researchers distributed the iPSC-derived hepatocytes into thousands of tissue-culture wells, added a different drug to each, and then measured whether the drug's presence led to a change in apoB levels. Among the "hits" recovered in this drug screen was a class of drugs known as "cardiac glycosides"—drugs like digoxin and digitoxin that are used to treat heart disease. Consulting an anonymized database of information regarding patients who had seen a doctor for heart disease, Duncan and his colleagues found that LDL levels in patients taking cardiac glycosides were just as low as they were in patients taking statins, drugs that are widely used for the specific purpose of lowering serum cholesterol. The cholesterol-lowering potential of cardiac glycosides—drugs that have been used clinically for more than two centuries—became apparent only with iPSC.

A final example of the promising applications of iPSC technology comes from the world of infectious disease. Hepatitis C virus (HCV), a leading cause of cirrhosis and liver transplantation in the United States, infects hepatocytes and causes a chronic immune reaction that slowly degrades liver function over time. Some HCV-infected patients develop severe disease, progressing to liver failure, while others have a benign course. But because the virus infects only human and chimpanzee hepatocytes, there is no mouse model of the disease, making the reasons behind this variability in outcomes difficult to study. iPSCs have provided a way in, as iPSC-derived hepatocytes can be infected with the HCV virus, making it possible to study the ensuing cellular and molecular misbehavior. And because the iPSC-derived hepatocytes can come from a variety of individuals, the approach allows researchers to study differences in HCV biology across the human population. More recently, this strategy has been used to study other viral pathogens that affect humans, from Zika and West Nile virus to HIV and SARS-CoV-2, the pathogen that causes COVID.

These are all first-pass approaches. The cellular alchemy of induced pluripotency has been around for less than two decades, and one can only imagine where the field will be two decades from now. Will each of us have our own iPSC line—an in vitro incarnation with which to further individualize our medical care? Will we reach a point where artificial organs can be made from a given patient's iPSCs, rescuing failing livers, kidneys, hearts, and even brains? These are possibilities, not inevitabilities. But one thing is certain: these tools will change the practice of medicine in *some* way.

In a sense, disease represents an undoing of the work of embryogenesis, a molecular deconstruction of the cellular networks built up during development. The chapters that follow will provide a glimpse into that process of undoing, and how the science borne out of the embryo may be able to slow, halt, or even reverse our decline. We'll begin where the field of embryonic stem cells did—the scourge of cancer, a disease that kills nearly 10 million people across the globe each year.

Chapter 9

ONE CELL RUN AMOK

Cancer didn't bring me to my knees,
it brought me to my feet.

—MICHAEL DOUGLAS

Only entropy comes easy.

—ATTRIBUTED TO ANTON CHEKHOV

At first glance, cancer would seem to bear little relationship to embryonic development. Tumors are disorganized and unpredictable, obliterating whatever tissues stand in the way of their inexorable effort to expand, while embryos are exemplars of organization and reproducibility. Even teratomas, the tumors that Roy Stevens used to model cellular differentiation, grow with unbridled zest, their internal structures having only the faintest resemblance to a normal tissue. Cancer is an act of destruction, while development is an act of creation.

But tumors and embryos have more in common than first meets the eye. The most obvious similarity is growth, a hallmark of both cancer and embryonic development. But the resemblances extend far beyond cell proliferation. Vast networks of cellular communication make tumors more than a simple collection of cancer cells. Programs of cell differentiation and plasticity, many of them carbon copies of procedures first established in the embryo, enable tumor cells to move and metastasize. Even stem cells may play a role in the complex ecology of a tumor.

The biggest differences between a tumor cell and its normal counterpart can be found in the genome. Theodor Boveri, the German developmental biologist who first discovered that chromosomes are the carriers of hereditary information, was also the first to provide a molecular explanation for cancer. Although Boveri never worked on the disease, his studies of embryos suggested strong links between development and cancer. The approach he used—making sea urchins with too many or too few chromosomes—had resulted in developmental abnormalities, malformed embryos that resembled tumors. Inspired by the observation that cancers exhibited altered chromosomes, Boveri put these pieces together, concluding that chromosomal disturbances are not merely correlated with cancer, they cause the disease. In 1914, Boveri codified his ideas into a simple theory of cancer. The model, which remains largely intact today, offered four concepts: (1) cancer is a disease of the cell; (2) cancers arise from one cell; (3) the tumor and its tendency to proliferate are caused by a chromosomal imbalance; and (4) the abnormal chromosomes, and the tendency to proliferate, are passed to the cancer cell's offspring.

We now know that cancer is a disease of the genes, and we have developed a deep understanding of how those genes (and the chromosomes they reside on) go awry during the formation of a tumor. Some genes promote cancer by pushing cells to grow and divide—these are known as "oncogenes." Other genes inhibit cell growth and division—these are known as "tumor suppressor genes." In their normal form, these genes are harmless and, indeed, necessary for embryonic and adult life. This duality of function—that these cancer-related genes can serve both benevolent and malevolent purposes—highlights one of the central principles of cancer biology: *tumors do not invent new biology, but instead use existing biology in new ways.*

During development, when rates of cell division are equivalent to those achieved in tumors, oncogene function dominates, and the embryo grows. Later, when development is complete and growth is no longer needed, the balance of gene activity shifts to tumor suppressor genes, and growth ceases. This regulated and dynamic balance between growth-stimulating and growth-inhibiting programs—calibrated over millions of years by evolution—allows animals to reach their predetermined size.

In cancer, mutations abolish this balance. The human genome contains dozens of oncogenes, and they promote cell growth and division in a variety of ways. Some, like the receptor for epidermal growth factor (EGFR), transmit growth-promoting signals from the outside world to the cell's interior. Others, like the MYC oncogene, are transcription factors, regulating the expression of hundreds of other genes that allow a cell to grow. Similarly, the human genome contains dozens of tumor suppressor genes, and they too act through a variety of molecular mechanisms.

You can think of a dividing cell as a car in motion, with oncogenes serving as the accelerator and tumor suppressor genes as the brakes. Normally, a driver controls a car's velocity by judiciously switching between the two, stepping on the gas to speed up and on the brakes to slow down. Cells do the same thing, alternating between proliferation and "quiescence," a nondividing state. In embryos, tissue expansion is the priority, placing cells on a growth highway—no brakes, and pedal to the floor. In the adult, with some exceptions, tissue maintenance is the order of the day, with most cells either parked or traveling at unbearably slow speeds.

Cancer-causing mutations perturb these functions, either by supercharging oncogenes (the equivalent of a brick on the gas pedal) or disrupting tumor suppressor genes (the equivalent of cutting the brake lines). Cancer is the cellular equivalent of a runaway car, and nature has imposed several barriers to prevent it from happening. Tumors emerge only when these fail-safe mechanisms don't work, when multiple mutations, or "hits," accumulate within the same cell. The exact number of hits needed to convert a normal cell to a cancerous one varies widely, but a single mutation is insufficient. Consequently, most tumors develop over years, starting out as "premalignant" growths—polyps in the colon or moles on the skin that become malignant only as more and more mutational hits accrue. As a result, the risk of cancer increases with age, as cancer-causing mutations take time to accumulate. Nature has safeguarded us against cancer in our reproductive years, fulfilling an evolutionary imperative, with little regard to what happens after.

THE CELLULAR ORIGINS OF CANCER

TOLSTOY FAMOUSLY WROTE THAT ALL HAPPY FAMILIES ARE ALIKE but every unhappy family is unhappy in its own way. The same is true of cancer, where each tumor is, in essence, its own species. There is no single molecular path to malignancy. While mutations in oncogenes and tumor suppressor genes are the sine qua non of cancer, each cancer is the result of a different combination of genetic alterations. As cells acquire these oncogenic hits, their behavior becomes increasingly erratic, a process referred to as "tumor evolution." And because tumors can evolve from many combinations of oncogene and tumor suppressor gene abnormalities, no two cancers are the same.

Every tumor starts from one cell. But unlike an embryo, whose starting place is apparent (the zygote), the cellular origins of a tumor are less obvious. A high-level view of cancer considers the tissue in which a malignancy arises as its source, leading us to refer to a tumor as "breast cancer" or "lung cancer." But at a more refined level, we know that each tumor comes from a specific cell, one of the millions present in each organ, and it is at this point that our understanding gets fuzzy. The lining of the colon, for example, contains a vast cellular array—differentiated absorptive and secretory cells as well as the stem cells from which they are derived. The question, then, is whether all these cells can serve as substrates for cancer, or whether that is a property of a select subset. And if tumors can have more than one cell of origin, does that make a difference biologically or clinically?

Defining the specific cell from which a tumor arises is daunting. Doing so properly would require a way to record a tumor's growth, from beginning to end, and then play the movie backward to the source. As no such recording ability exists, and a tumor's parentage cannot be determined by its appearance alone, we have instead relied on genetically engineered mouse models of cancer.

The first of these models emerged in the 1980s as a product of *transgenic mouse* technology—the ability to create mice carrying foreign pieces of DNA (oncogenes) in their genomes. These early models were unambiguous proof that oncogenes cause tumors, and they offered methods for studying cancer in living animals. Later, researchers applied the mouse "knockout" technology pioneered by Evans, Capecchi, and Smithies to cancer, allowing them to develop more sophisticated models of tumors—models on which cancer researchers have pinned their hopes for new and better treatments.

IN MY LABORATORY, WE RELY ON ONE OF THESE MODELS TO STUDY pancreatic cancer, a highly lethal disease for which we lack adequate treatment options. The model was created in the early 2000s by David Tuveson and Sunil Hingorani, and it relies on a pair of genes that are commonly mutated in pancreatic tumors—the KRAS oncogene and the p53 tumor suppressor. Both genes are mutated in many types of cancer, but their involvement in cancer of the pancreas is especially common. Mice carrying these mutations are referred to in the field as KPC mice—"K" stands for KRAS, "P" stands for p53, and "C" stands for a molecule called Cre that converts the normal alleles of these genes into their cancer-causing counterparts. By design, these mutations are directed specifically and exclusively to the pancreas of KPC mice, but they are insufficient on their own to cause tumors to form. Other mutations accrue gradually (in keeping with the "multiple hit" paradigm of cancer), and it isn't until two to three months of age that "precancerous" lesions are visible under the microscope. Over the next few months, however, a threshold of genetic and epigenetic events is crossed, and KPC animals develop metastatic pancreatic cancers, simulating the damaging reality of the human disease.

Similar genetically engineered models have been created for nearly every tumor type. In addition to their utility in testing possible therapies—immunotherapy drugs, agents that interfere with tumor metabolism, or those that can overcome therapy resistance—genetic models of cancer have also been used to address the question of cellular origins. Some studies indicate that tissue stem cells are especially prone to cancer. For example, introducing cancer-causing mutations into intestinal stem cells initiates the growth of tumors, whereas introducing the same mutations into cells that have already differentiated has a more limited effect. But in other tissues, especially those in which stem cells may or may not exist, tumors seem capable of arising from multiple types of cells. Current thinking in the field holds that most cells in the body can give rise to cancer, but their individual propensities vary from one cell type to the next.

There are two exceptions: (1) neurons, the cells that carry electrical impulses in the brain, spinal cord, and peripheral nervous system, and (2) cardiomyocytes, the muscle cells that give the heart its pump. These two cell types appear to be impervious to cancer, and it is likely no coincidence that these are among the only cells in the body that have lost the ability to divide. This constraint on division can be devastating, for it means that the neurons or cardiomyocytes that die following a stroke or heart attack cannot be easily replaced. The silver lining of this limitation is that they are protected from forming tumors, which perhaps holds some lesson for how we might prevent other cancers in the future.

PHYSICIANS AND SCIENTISTS CATEGORIZE CANCERS IN VARIOUS WAYS. The first level of description is the organ in which a tumor first appears—lung cancer, breast cancer, and so on. Layered beneath this is the tumor's appearance under the microscope—its "histopathology." Pathologists classify tumors according to the cellular structures they observe; "carcinomas" (the most common and lethal malignancies affecting humans) are defined by their epithelial features, "sarcomas" (Greek for "flesh")

by their mesenchymal features, and "lymphomas" by their microscopic resemblance to lymphocytes and their concentration within lymph nodes.

These are the classical ways of grouping tumors. But over the past few decades, a new categorization scheme has seized the spotlight. This classification system relies on a tumor's molecular features—the specific set of DNA mutations, chromosomal aberrations, and gene-expression patterns that constitute the driving force behind the cancer's malignant growth. Treating tumors based on their molecular makeup, rather than tissue of origin or microscopic appearance, is the basis of "precision oncology," which tailors a treatment plan to the specific susceptibilities found in an individual tumor.

We observed earlier that because each tumor contains a different blend of molecular hits, there is no single path to cancer. The hundreds of oncogenes and tumor suppressor genes in our genomes could, in principle, lead to a vast number of cancer-causing combinations, but it turns out that not every combination has that capacity. Instead, each tumor type has its own mutational footprint. While most pancreatic cancers carry mutations in the KRAS and p53 genes, the prevalence of these and other mutations varies across cancers. Many colorectal cancers, for example, may harbor KRAS and p53 mutations, but most also possess mutations in a gene called APC, which is almost never mutated in pancreatic cancer. Chronic myelogenous leukemia—a cancer of the blood—rarely acquires mutations in any of these genes, but instead develops through mutations affecting the ABL oncogene.

While most cancer-causing mutations accumulate over the lifetime of an individual, some are inherited, serving as the basis for familial tumor syndromes. But *carriers* of such inherited mutations are at increased risk for only certain tumor types. For example, individuals who inherit a mutant copy of the BRCA1 or BRCA2 gene are at significantly increased risk of breast, ovarian, prostate, and pancreatic tumors, but not other tumor types.

All of this suggests that a tissue's developmental history defines the spectrum of mutations that can cause a tumor to grow. But why is this so?

Again, we can only speculate. The most important factor dictating a tumor's molecular features is its tissue ancestry, and this connection brings us back to the age-old dichotomy of nature versus nurture. A cancer's nature (its developmental origins) accounts for some of its features, while nurture (mutations caused by environmental insults like cigarette smoke, ultraviolet light, and toxic chemicals) accounts for others. These two components—*which* mutations occur in *which* cell type—likely converge to provide the most complete picture of a tumor's malignant trajectory and its vulnerabilities. We are well equipped to find mutations, as advances in DNA sequencing allow us to find them easily, whether we are surveying thousands of tumors or an individual patient. It is our ability to identify the source of cancer that is lagging—to know the nature of the cell from which a given tumor began.

Each tumor has its own unique origin story, and therein lies its Achilles' heel.

DE-DIFFERENTIATION AND PLASTICITY

IF DIFFERENTIATION PROVIDES A MEASURE OF DEVELOPMENTAL PROgression, then the opposite process—de-differentiation—marks the progress of a tumor. Like cellular Benjamin Buttons, cancer cells lose their specialized features the older they get, ultimately attaining the undifferentiated characteristics of their embryonic forebears.

If we view cancer as the product of multiple hits, then each of those molecular blows prods the affected cells closer and closer to malignancy. With each new assault on the genome, cells gain unwelcome attributes—an ability to grow and spread beyond the confines of their normal position in a tissue. But paralleling the acquisition of new traits is the loss of old ones, for cells are forced to abandon their places in the cellular society when they embark on the path to malignancy.

This sequence of events is easily tracked in the colon, where colonoscopy provides an opportunity to see how precancerous polyps can progress to cancer. The first step in this malignant progression is the emergence of an "adenoma," a region of tissue where cells have lost the normal, differentiated features of their neighbors. With further genomic hits, the cells lining the colon's epithelium become progressively less differentiated (what pathologists refer to as varying degrees of "dysplasia"), bearing features that resemble embryonic rather than adult intestinal cells. Finally, the cells no longer remain in their designated positions, and begin pushing their way deeper into the tissue or slipping into a blood vessel. At this point, the growth is considered a cancer.

Is de-differentiation required for a cancer to form? Or is the reversion to a more primitive state merely an epiphenomenon, a result of the cancer cells turning their attention elsewhere? There is evidence on both sides, but some studies suggest that the differentiated state is incompatible with cancer. The clearest example comes from a type of blood cancer called acute promyelocytic leukemia (APML), which is caused by a chromosomal "translocation" that joins together two chromosomes that are normally separate—chromosomes 13 and 15. APML cells have a less differentiated appearance than their normal counterparts in the blood, neutrophils. But exposure to all-trans retinoic acid (ATRA)—an embry-

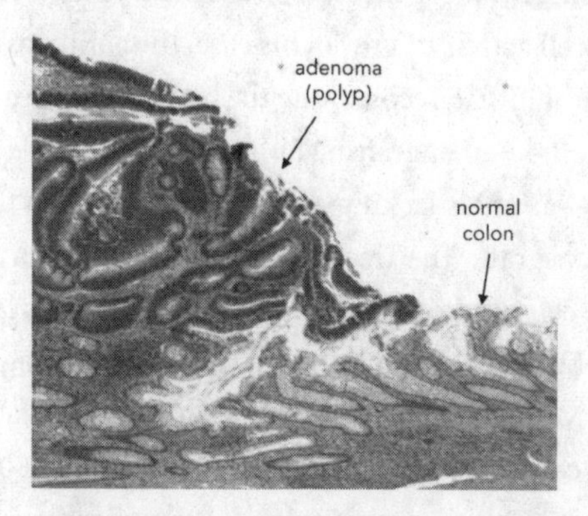

The beginnings of cancer. Viewing them through a microscope, it is easy to tell the difference between the epithelial cells that line the colon normally (lower right) and those that have taken the first step toward cancer as a premalignant polyp, or "adenoma" (upper left). Colon polyps contain one or more mutations–the "hits" that set the stage for cancer to develop.

(IMAGE COURTESY EMMA E. FURTH, MD.)

onic signaling molecule—initiates a differentiation program, causing the leukemia cells to turn into harmless neutrophils. Unlike chemotherapy, which is designed to kill rapidly dividing cells, ATRA doesn't execute its targets; instead, it reminds them of their potential to return to normalcy, converting their life-threatening swords into ploughshares.

This type of treatment is known as "differentiation therapy." The discovery of ATRA was serendipitous, making APML one of the few successful applications of differentiation therapy to date. But this exciting approach remains under investigation for other types of tumors. How fitting it would be if the signals that drive cells to mature in the embryo could cajole cancer cells into doing the same thing!

IN 1997, CANADIAN SCIENTIST JOHN DICK PUT FORTH THE CLAIM THAT cancer cells exist within a cellular hierarchy, akin to the hematopoietic stem cell hierarchy that James Till and Ernest McCulloch, his predecessors at the Ontario Cancer Institute, had described for the blood 30 years earlier. Like his prior mentors, Dick used cellular transplantation to assess stem cell activity. Only, in his case, the ability to form a cancer, rather than the ability to reconstitute the blood, was the assay. Dick first developed a method for engrafting human leukemia cells into immunodeficient mice—animals lacking an immune system so that they could accept a human graft. The transplants took only when he used more than a million cells, indicating that most leukemia cells lacked the ability to form a tumor. Next, by diluting the cells (taking another page from Till and McCulloch's playbook) he determined that these "leukemia-initiating cells" were extremely rare. Out of every 250,000 leukemia cells, only one was able to reconstitute the leukemia in a recipient mouse.

To identify those needle-in-a-haystack cells, Dick divided the leukemic cells according to the presence or absence of two proteins on their surface—markers bearing the names CD34 and CD38. After sorting the cells into their four possible subpopulations (positive for both markers, negative for both markers, or positive for one or the other), he tested each for its leukemia-initiating potential. Only cells that were positive for CD34 but negative for CD38 grew into leukemias. The conclusion was simple enough: the ability to spawn a new leukemia resided solely within this population of cells.

The "*cancer stem cell* hypothesis," as John Dick's theory came to be called, views tumors as a composite mixture of two types of cancer cells—stem cells (rare cells responsible for fueling the tumor's growth) and non-stem cells (the bulk of the tumor lacking the ability to self-renew). It bears a striking similarity to the model that Roy Stevens and Barry Pierce conjured up from their work on teratomas, which found that the cells of a teratoma took on either a differentiated or an undifferentiated appearance, and that only the latter could form new teratomas. Over the past two decades, researchers have used similar transplantation strategies to identify putative cancer stem cells in virtually every type of solid tumor, leukemia, or lymphoma.

In its most extreme version, the cancer stem cell hypothesis posits that all of a tumor's growth potential is contained within a small subset of cells—the cancer stem cells. If correct, the idea has important implications for medicine. Normally, physicians judge the efficacy of anticancer therapies based on their ability to reduce the overall size of a tumor. The best therapies, therefore, are judged to be those that shrink tumors the furthest. But even those therapies that can dramatically shrink a tumor rarely cure the cancer, and the cancer stem cell hypothesis offers an explanation for why this might be: a tumor forced into remission with standard therapy—drugs that kill off most of the tumor cells without affecting the cancer stem cells—would be destined to grow back, as any remaining stem cells would provide the substrate for regrowth.

By contrast, a drug that exclusively killed the cancer stem cells but had no effect on the rest of the tumor cells might be the ultimate cancer therapy, as it would in effect poison the cancer's root system. Were such a therapy to be found one day, it would behave very unusually. At first, the drug would appear to have minimal effects, as its cellular targets—the cancer stem cells—make up only a small fraction of the total tumor mass. But over time, as the bulk of the tumor cells succumbed to attrition and failed to be replaced, the tumor would melt away like a plant whose root system was poisoned.

The CSC hypothesis is controversial, and its alternative is the simple arrangement wherein most cells in a tumor, not just a select subset, have the capacity to divide. If this is the case, how do we explain the results of Dick and others who showed that only a chosen minority of tumor cells have tumor-initiating activity? One possibility is that the assay itself is misleading (an omnipresent danger in laboratory research). While transplantation is a useful way to identify cells that can initiate tumors when transplanted in a laboratory setting, it may not accurately reflect what is happening in an actual patient.

Alternatively, in the same way that some (normal) organs are maintained by stem cells while others are maintained by uncomplicated cell divisions, some tumors may indeed operate through a CSC-driven cellular hierarchy while others may employ a more egalitarian cellular order. Finally, even if CSCs are responsible for fueling the growth of a tumor, it is hard to imagine how to develop therapies that would target them selectively. As attractive a model as it is, the CSC hypothesis will require further proof before it is accepted, much less used in a clinical setting.

EARLIER IN THE BOOK, IN OUR EXPLORATION OF THE PROCESS BY which the embryo takes shape, we touched on the concept of epithelial plasticity—the shape-shifting changes in cellular phenotype that embryonic cells use to acquire form. The phenomenon makes its most dramatic appearance during gastrulation, the life stage that Lewis Wolpert quipped is more important than birth, death, or marriage. This spectacle—in which the cells of the single-layered epiblast are realigned into a tri-layered embryo—is achieved through a form of cellular plasticity in which the epiblast's cells lose the epithelial features that bond them tightly to their neighbors, known as an epithelial-to-mesenchymal transition, or EMT. This transition occurs repeatedly throughout development, as does its opposite, MET.

In normal tissues, cellular plasticity of either type—EMT or MET—ceases once embryogenesis is complete. But in cancer, tumor cells dredge up their dormant migratory abilities, remnants of the form-building programs their embryonic ancestors used for morphogenesis. When cells lose their epithelial ties, they acquire the migratory properties of mesenchymal cells, enabling them to break their epithelial bonds. The consequence is metastasis, the dissemination of cancer cells to other parts of the body.

The presence or absence of metastasis is the most important feature determining a patient's prognosis. Before cancer spreads, surgery can often be curative. But once its cells have disseminated to other parts of the body, a tumor's potential to do harm grows exponentially. Metastatic tumors can seed infections, predispose patients to clots, and cause organs to fail, making metastasis the most common cause of death associated with cancer. This is, of course, just the clinical way of looking at the process. On the human side, metastasis is a formula for fatigue, pain, and misery.

Paradoxically, metastasis is an inefficient process. To escape the primary site and establish a new home, a cancer cell must run a gauntlet of challenges, collectively referred to as the "metastatic cascade." In carcinomas, the first step is *invasion*, whereby cancer cells disconnect from their neighbors and break through the *basement membrane*, the collagen-rich carpet on which epithelial cells normally rest. It is here that EMT plays its most important role in metastasis, for as cells are freed from their epithelial bondage, they traverse the basement membrane and make their way into deeper structures. Next, cancer cells must *intravasate* into a blood vessel or lymphatic channel, giving them access to the body's transit systems. Finally, the cells must then leave the vascular system and be able to grow in, or "colonize," the secondary site. But each of these steps occurs infrequently. Consequently, fewer than 0.00001 percent of the billions of cells present in a primary tumor have any chance of forming a metastasis.

Beyond facilitating the first step in metastasis, EMT is also a means by which cancer cells become resistant to drugs. There are many forms of cancer therapy, each with its own mechanism of action, but almost all work by depriving rapidly dividing cells of the signals or building blocks that they need to grow and survive. In response to the selective pressures imposed by anticancer therapy, cancers evolve to escape the toxic effects of drugs. In some cases, they achieve this through mutation—a Darwinian-like process whereby mutations that improve fitness (that is, resistance to the drug) lead the cells carrying them to be enriched and expand, as with the emergence of a new "species." But in other cases, cancers gain resistance without mutation, through cellular plasticity. EMT is a common pathway to resistance because mesenchymal cells often have different drug sensitivities from their epithelial counterparts.

The similarities between embryos and tumors present rich opportunities for new anticancer strategies. But first, we must bridge several large gaps in our knowledge. We have yet to figure out how to treat tumors other than APML by forcing cancer cells to differentiate into more benign forms. We still struggle to find cancer stem cells in living tumors and wouldn't know how to target them even if we could. Finally, we do not understand what provokes cancer cells to undergo EMT, and we have no drugs with which to block the process.

Cancer, as we knew all along, is quite a complicated adversary.

TUMOR NEIGHBORHOODS

WE HAVE SPENT A LOT OF TIME TALKING ABOUT THE CANCER CELL—its origins and the events that govern its uncontrolled growth. But it may surprise you to learn that in many tumors, cancer cells make up a minority of the tumor mass. Most of the cells in these tumors are not cancer cells at all but, rather, normal cells that the tumor has trapped as indentured servants—immune cells, blood vessels, and fibroblasts that serve the tumor in some nefarious way. Fewer than a fifth of the cells present in a tumor may be cancer cells. These noncancerous elements create cellular neighborhoods referred to collectively as the *tumor microenvironment*, and their existence points to new avenues for research and treatment.

An indispensable part of a tumor's noncancerous component is its blood supply, which provides sole access to the nutrients and oxygen that are essential for the tumor's growth. In 1971, Harvard surgeon Judah Folkman asserted that cancers must recruit new blood vessels to feed their insatiable appetite for raw materials, a claim that initially met with skepticism. But when President Nixon declared his "war on cancer," the idea moved from the fringes to the mainstream, earning Folkman, and Harvard, huge industry grants.

The growth of new blood vessels is known as *angiogenesis* (from the Greek *angio-*, meaning vessel, and *genesis*, meaning birth). The phenomenon was first described by the eighteenth-century British surgeon John Hunter, who noticed new blood vessels sprouting in the antlers of reindeer after winter, and we have since come to realize that it is critical to a variety of physiological and pathological conditions. Angiogenesis is crucial during development, as blood vessels expand to feed the growing embryo with nutrients, but it plays a minor role in adult animals—at least under normal conditions.

Folkman's insight was that tumors would quickly outgrow their blood supply if the cancer cells failed to reengage angiogenesis programs. He then reasoned that any drugs that inhibited the process could turn out to be the perfect anticancer therapy—a silver bullet—since angiogenesis was something that cancer cells needed but normal cells, with their vascular system already established, did not. If a treatment could stop angiogenesis, the tumor would starve. Over the following two decades, Folkman and others set out to identify the molecules responsible for angiogenesis. They found that many of the capillary-inducing signals came from the cancer cells themselves—latent programs summoned up from embryonic life that served as beacons to recruit new blood vessels.

The signaling molecule responsible for much of the angiogenesis in cancer is a secreted protein called vascular endothelial growth factor (VEGF). The protein and the gene encoding it were isolated in 1989 by a team at Genentech led by Napoleone Ferrara, and over the next decade the company worked to develop an inhibitor that could block the molecule's ability to stimulate the growth of new blood vessels. In 2004, the Food and Drug Administration granted the biotech giant approval to use the company's VEGF-blocking antibody, Avastin, to treat patients with colorectal cancer. Since then, the drug has also been approved to treat many other cancer types, including lung, kidney, and brain cancers.

Unfortunately, Folkman's vision (he died in 2008) has been only partially fulfilled. Anti-angiogenesis therapies form an important part of the modern oncology arsenal, but they have not had the near-universal impact he anticipated. Underlying these failures is the "redundancy" of biological systems; when VEGF is blocked, it turns out, other signals can take over to mediate angiogenesis. Redundancy can have benefits in normal tissues, providing cells with a backup system if a signaling pathway misfires. But in tumors, redundancy is a blight, for it gives cancer cells a way to bypass the effects of anticancer drugs.

Beyond regulating their own blood supply, cancer cells shape their cellular neighborhoods in other ways, again utilizing developmental signals to force normal cells to do their bidding. One important component of the tumor microenvironment is immune cells, whose principal job is to detect and eradicate microbial invaders. (During embryonic development, the immune system learns which chemical structures are a normal part of the body, causing it to consider as "foreign" any substances it encounters later.) But the immune system also moonlights as an anticancer agent. As tumors grow, their cells inevitably generate new chemical structures. The immune system, in turn, recognizes these structures as foreign and eradicates the nascent cancer cells that produced them.

But such immune protections can be short-lived. As we learned earlier, cancer is an evolutionary process. Faced with the selective pressure of immune attack, cancer cells devise various ways to evade recognition.

For example, they may recruit cells that constitute the protective arm of the immune system, dampening its antitumor activity. Cancer cells may also induce the expression of "immune checkpoint" molecules—proteins that act as a brake on immune attack, thereby shielding the cancer cells from annihilation. Drugs that block the activity of these immune checkpoints unleash the immune system's powerful antitumor activity, and they have already revolutionized cancer care in their first decade of clinical use.

Finally, cancer cells also recruit fibroblasts, the connective tissue cells that play a key role in wound healing, again using the same signals employed during development. Fibroblasts provide cancer cells with metabolites, growth factors, and other rewards. But as my team and others showed several years ago, cancer-associated fibroblasts come in both tumor-promoting and tumor-suppressing flavors. Leveraging the antitumor activities of fibroblasts while simultaneously suppressing their protumor activities has become one of the pressing areas of contemporary cancer research.

CANCER CELL AS ORGANIZER

IT HAS BEEN A CENTURY SINCE HANS SPEMANN AND HILDE MANGOLD identified the embryonic organizer as a group of cells that held dominion over its neighbors, "organizing" their identities, movements, and fates. Since then, we have come to understand that tumors are complex tissues that owe their malignant potential to molecular programs co-opted from embryonic development. With their altered genomes, cancer cells hold dominion over the normal cells that form the tumor microenvironment, yet they are still dependent on these cellular subordinates.

Perhaps not surprisingly, the channels of communication within a tumor are the same ones used by embryos—evolutionarily conserved signaling pathways we encountered in previous chapters, like notch, wingless, and hedgehog. Cellular cross talk is universal in cancer, just as it is with embryos, as conversations transpire among all parties present.

The more we learn about these cellular exchanges—which are easier to overhear in the more structured setting of the embryo—the greater our ability to amplify or squelch them for therapy.

The level of complexity within a tumor is daunting, but all is not chaos. If we consider the early stages of tumor formation, the time when the tumor microenvironment is just forming, there is a chain of command. A fibroblast may have a discussion with an immune cell, and an immune cell may converse with a vascular cell, but all of them will answer to the same commanding officer: the tumor cell. In that sense, cancer cells function as the pernicious organizers of a tumor, with the power to reshape its surroundings to best advantage.

Biologists have long recognized the parallels between tumors and embryos, and Boveri introduced the idea that differences in the genome are what separate the two. The war on cancer (if "war" is the proper metaphor) began without precision weapons—relying on surgery, radiation, and toxic chemicals to eradicate or incapacitate the tumor. Subsequently, in the late twentieth century, oncologists started to rely increasingly on molecular and genetic information from the tumor cells to create *targeted* therapies—medicinal smart-bombs that would home in on the susceptibilities conferred by a tumor's unique biochemical features. We are now in the third phase of our biomedical conflict with cancer, one that recognizes tumors as more than a collection of cancer cells, but rather as the evil doppelgängers of the embryonic tissues from which they arose.

The embryo is likely to hold even more secrets that will help scientists and physicians better diagnose and treat cancer. But it is also possible, as the deep connections between cancer and development are better understood, that cancers may be able to teach us something about the embryo.

Chapter 10

EYE OF NEWT AND TOE OF FROG

Every component of the organism is as much of an organism as every other part.

— BARBARA MCCLINTOCK, *A FEELING FOR THE ORGANISM*

If there were no regeneration, there could be no life. If everything regenerated, there would be no death.

— RICHARD GOSS, *PRINCIPLES OF REGENERATION*

Karen Miner, a 41-year-old real estate agent, was halfway back to her home in California wine country when she had the accident that would leave her paralyzed. Miner's clients were having problems with their lender ("It was an escrow from hell," she remembers), and she agreed to make the hour-long drive from her house to theirs so they could work through the paperwork together. Miner often brought her daughters—aged 4 and 5—along with her to work, but for this longer-than-usual trip she left them with a sitter.

"We had it all," she recalled. "I was so happy; I was afraid it couldn't last."

It started to rain as Miner got underway—the first storm of the fall rainy season. She took extra care as she maneuvered through the winding hills, gently applying the brakes with each twist in the road as she had been taught to do in wet conditions. But one curve came a little too fast, and her Jeep careened through the guardrail, rolling side over side down a hill before settling in the ravine below. Miner was conscious the whole time. She realized that she felt no pain—the first sign that something was seriously wrong—and when she reached for the keys to turn off the engine, her hands failed her. Then, slowly, and surreally, her head slumped to her chest, where it came to rest in an unnatural position. She tried to move but couldn't. A passing motorist stopped to help and stabilized her head and neck as the rain pelted them both. After what seemed like an eternity, an ambulance arrived, and Miner's world went black.

ROUGHLY 300,000 PEOPLE IN THE UNITED STATES LIVE WITH A DISability related to spinal cord injury (SCI). Motor vehicle or occupational accidents are the most common causes; a smaller fraction result from falls, gunshot wounds, sports injuries, and other mishaps. An analysis published in 1998 estimated that the annual cost of care for patients with spinal injuries approached $10 billion. But this outdated number captures only a fraction of the total cost to society, as individuals with paralysis end up relying on friends and family for their care—services that most likely exceed hundreds of billions of dollars of unreimbursed work each year.

The nerves that run through the spinal cord are divided in space according to their purpose; those controlling motor function are positioned in the front of the cord while those controlling sensation are in the back. Nerves sense or stimulate different parts of the body depending on the level at which they exit the vertebral column. Consequently, the deficits associated with SCI depend on the location and extent of cord damage. Injuries involving the upper neck are the most dangerous, as this is where the nerves that control breathing reside. Injuries further down the spine are associated with motor and sensory deficits, again depending on the level of injury. Bowel and bladder dysfunction occur with almost all SCIs, since the nerves that regulate sphincter control exit the cord at the base of the vertebral column.

After a spinal cord injury, it is common for patients to recover a modest degree of function. This occurs via "sprouting," a phenomenon in which surviving nerves extend fresh branches ("axons") in search of new neuronal connections (synapses). Sprouting is most effective in the peripheral nervous system—the nerves that lie outside of the brain and spinal cord—where nerves can sprout with amazing speed. (Neurons in the arm or the leg can extend axons at rates exceeding an inch per month.)

By contrast, injuries involving the central nervous system—the brain and spinal cord—are commonly associated with scarring, the body's vigorous but often misguided effort to heal itself. But scar tissue, which impedes sprouting, is only part of the problem. A second and even greater hindrance to recovery is the fact that the nervous system is incapable of enlisting new recruits. While most cells in the body can divide in response to an injury, neurons are different. As we saw in the last chapter, neurons rarely if ever divide after birth, and while this makes them resistant to cancer, it also prevents them from regenerating.

For all intents and purposes, we are born with all the neurons we will ever have.

WHEN SHE AWOKE IN THE HOSPITAL WITH A BREATHING TUBE IN HER trachea, Karen Miner had no idea where she was. The sedatives that were supposed to make it easier for her to breathe also caused hallucinations, and it was several days before she became oriented to her surroundings. Miner had been taken to Santa Clara Valley Medical Center, a federally designated spinal cord injury center, where neurosurgeons removed residual debris pressing on the cord. They also administered intravenous corticosteroids—anti-inflammatory drugs intended to reduce swelling that might further compromise function.

The accident had damaged the upper portion of Miner's neck,

involving the third and fourth cervical vertebrae (abbreviated C3-C4). But the injury was incomplete, primarily affecting the nerves running through the middle of her spinal cord, thus sparing her breathing. This type of injury results in what doctors call an "inverse paraplegia," in which the arms and hands are more severely affected than legs and feet. In Miner's case, this meant that her upper arms were locked in position, a consequence of muscle stiffness or spasticity, even though she could stand and bear weight.

After months of intense physical therapy, Miner finally went home, returning to a life that was unrecognizable. Home-health aides became a rotating fixture in her house, providing 24-hour assistance. But it was her daughters who became her most important caregivers. Early on, they learned how to place the urinary catheters that she needed to empty her bladder, becoming so adept at the maneuver that they began teaching the aides how to do it effectively. Burdening her daughters was hard enough, but the worst part was the constant struggle for independence. "Having caretakers there was the hardest thing of all, harder than anything else," Miner recalled.

Compared to many others with SCI, Miner managed well. She had the support of friends and family and was lucky enough to be financially secure. Over time, she regained more and more independence. But for many individuals, the loss of function and associated pain are unbearable, and suicide accounts for up to 10 percent of deaths following SCI. (Other leading causes of death for individuals with SCI are infections and blood clots.)

The actor Christopher Reeve became the most famous spokesperson for SCI in 1995, when a horsing accident resulted in a high vertebral fracture (C1-C2). Paralyzed from the neck down, Reeve became dependent on a ventilator for the rest of his life, able to breathe on his own for sporadic intervals lasting at most 90 minutes at a time. Hopes were high when, after several years of intensive physical therapy, Reeve regained the ability to move his fingers and sense touch in parts of his body. But the gains did not herald further improvements, nor did they prevent his death in 2004 from sepsis, the result of a pressure ulcer associated with his immobility.

ORGAN FAILURE AND REGENERATION

THE WORD "ORGAN" IS DERIVED FROM THE GREEK *ORGANON*—literally "tool" or "instrument"—referring to a method for logical inquiry or philosophy (Aristotle's six books on logic are collectively referred to as the Organon). In the Middle Ages, it was a generic term for wind-based instruments, before it came to mean the large pipe-based devices we are now familiar with. (The word's corporeal meaning—as in "internal organs"—emerged during the fourteenth century.) The organist creates sound by adjusting numerous mechanical contrivances—keyboards, foot pedals, stops, and couplers—to control the flow of air through hundreds or even thousands of pipes, each with its own pitch and volume. Between the seventeenth and nineteenth centuries, the pipe organ was arguably the most complex device to be built on earth. With a dynamic range spanning the slightest *pianissimo* to the grandest *sforzando*, it is no wonder Mozart called the pipe organ the "king of instruments."

The musical instrument, like its anatomical counterparts, can fail in any number of ways. Acute malfunctions—a ruptured bellows or a fractured pedal—tend to announce themselves with great fanfare, eliminating certain notes or taking out entire registers. There is nothing subtle about acute failure.

Wear and tear, on the other hand, can easily go unnoticed. Accumulated dirt or rust in an organ's pipes may gradually throw off its richness and tone, but such changes can easily be ignored for long periods of time. Chronic failure happens out of sight, allowing the instrument to function as persistent damage crosses some critical point of no return.

Our internal organs are likewise prone to failure, and this can also occur in ways that are either overt and catastrophic or sluggish and destabilizing. When a tissue fails acutely—the result of blunt trauma, heart attack, or stroke, for example—it is hard to miss. Acute failures are medical emergencies, a rallying cry for health-care professionals. But in human beings, most organ damage is insidious, caused by corrosive diseases—autoimmunity, atherosclerosis, metabolic imbalances, or simple wear and tear—that sneak up on their victims, invisible unless someone goes looking. Sadly, we typically become aware of an organ's death throes from chronic failure only after it is too late to intervene.

GROWING UP IN THE EIGHTEENTH-CENTURY TOWN OF SCANDIANO IN northern Italy, Lazzaro Spallanzani had abundant opportunities to study the insects, lizards, and crayfish that populated the nearby woods. It is possible that the idea of studying organ regeneration first occurred to him there, perhaps manifested by amputating tails or claws of those resident creatures and observing the consequences. But the adult Spallanzani—priest, lawyer, and naturalist—proved to be more patient and persistent than his contemporaries or predecessors. By resolutely studying his experimental subjects over time, Spallanzani provided the first comprehensive description of animal regeneration.

"Regeneration" is a tricky term, carrying different meanings in different contexts. In mammals, cell proliferation is the most common response tissues make to injury (as noted above, neurons are the exception). This kind of recovery is known as *compensatory growth*, and it works well when the degree of damage is small. But if we turn our gaze to other branches of the tree of life, we see that nature has favored our evolutionary cousins with regenerative abilities we can only dream of—the ability to grow new arms, legs, heads, or tails, complete with nerve, muscle, blood vessels, and skin. This type of repair, in which a tissue is rebuilt with all its structural complexity intact, is called *epimorphic regeneration*. Mammals do not engage in epimorphosis, and thus we cannot, and perhaps never will, regrow organs in the fashion of our distant relatives.

The first people to witness epimorphic regeneration were men and women who earned their living from the sea. Seeing that the crustaceans they caught in their nets—principally crabs and lobsters—regularly bore a smaller-than-normal limb, these fisherfolk inferred that the animals had lost appendages, which had partially regrown. Scholars dismissed their anecdotes as folklore until 1712, when French naturalist René-Antoine Réaumur undertook a series of systematic amputations of freshwater crayfish. To his amazement, and that of any onlookers, whenever or wherever a limb was cut, a new one gradually grew back with extraordinary precision and accuracy.

An even more attention-grabbing discovery came a few decades later, when Swiss naturalist Abraham Trembley discovered that the hydra—an inch-long sea anemone–like sea creature—shared or exceeded this capacity for regeneration. When Trembley cut the creature in two, each half formed a new animal, complete in every respect. Trembley had stumbled upon a new form of reproduction.

Then, in the late 1700s, Spallanzani took these observations a step further, subjecting worms, tadpoles, snails, slugs, salamanders, toads, and frogs to the caprices of his scalpel. Despite their variation in size and shape, these organisms shared a remarkable capacity to regrow their severed tails, arms, jaws, and antennae, leaving no evidence of the injury they had suffered. But it was Spallanzani's discovery that decapitated snails could grow new heads that garnered the greatest attention. Once word got out, anyone in Europe who was interested in confirming the phenomenon with their own hands—scholar or commoner—could attempt a regeneration experiment. Even the French writer-philosopher Voltaire, having replicated the finding himself, proclaimed, "Not long ago everybody talked only about Jesuits, at present snails are the talk of the town."

Spallanzani's results revealed two important principles of regenerative biology. First, an organism's ability to regenerate is inversely correlated with its developmental or evolutionary state. Snails can regrow body parts that salamanders cannot, tadpoles regenerate more efficiently than adult frogs, and all of them do a better job than mammals. Second, animals regenerate only as much as they need to—no more, no less. A salamander regrows its limb whether it is amputated above or below the elbow, an indication that the tissue somehow "knows" the extent of injury that must be repaired. Likewise, regeneration occurs only from the point of amputation outward, so that an amputation below the elbow causes the forearm and fingers to regrow but leaves the upper arm and shoulder intact.

Since Spallanzani's time, many other examples of epimorphic regeneration have been discovered—animals that can regrow jaws, eyes, ovaries, and even parts of the spinal cord. In almost all cases, the process depends on the formation of a wound *blastema*, a specialized epithelial structure that grows to cover the amputated stump. The blastema is not

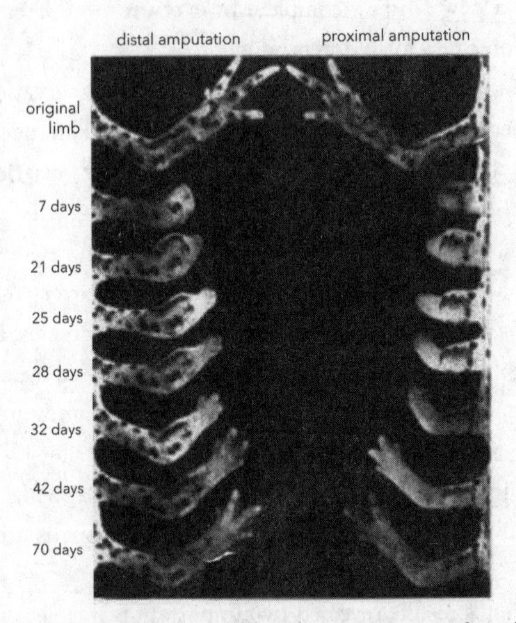

A salamander can regenerate an entire limb in a matter of weeks. Full recovery occurs whether the amputation occurs below the elbow (distal, left) or above the elbow (proximal, right), indicating that the tissue somehow "knows" exactly how much needs to be reconstituted.

(IMAGE REPRODUCED FROM *PRINCIPLES OF REGENERATION* BY RICHARD J. GOSS, WITH PERMISSION FROM ELSEVIER.)

simply a protective structure, however—it signals to the cells beneath, causing them to de-differentiate, a cellular state from which they can retrace their developmental paths to fuel regeneration.

But the regenerative powers we have discussed so far—those of amphibians and crustaceans—pale in comparison to the grand masters of regeneration, the planarian flatworms like *Schmidtea mediterranea*. Up to a half inch in length and bearing photoreceptors that eerily resemble eyes, these animals live in freshwater ponds, where they subsist on a diet of insects and larvae. So vast is their regenerative ability that an entire animal can regrow from 1/250th of a worm—the equivalent of a new person growing from a severed foot.

Planaria have a unique means of regeneration, for their bodies are filled with specialized stem cells called "neoblasts." After being cut, planaria form a wound blastema just like any other organism capable of regeneration. But in contrast to amphibians and crustaceans, which lack neoblasts and instead rely on de-differentiated cells, planaria use their ubiquitous stem cells to reconstitute form. These properties enable planaria to reproduce in the same fashion as Trembley's hydras: the worm anchors itself on a solid substrate and pulls itself apart, after which a new worm grows from each of the separated halves.

There is much we do not understand about regeneration, but one thing is clear: for a tissue to reclaim its normal form, the cells responsible for regeneration must recognize where they are and what they need to do. In other words, they must "know" where the amputation blade fell.

In the 1960s, biologist Lewis Wolpert presented a theory to explain how embryonic cells recognize their positions in a three-dimensional world. Wolpert's model posited that each cell acquires, in addition to its phenotypic or differentiated identity, a *positional identity*—a spatial designation that distinguishes each cell from others along dorsal-ventral (back-to-front), anterior-posterior (top-to-bottom), and proximal-distal (inner-outer) axes. Amphibian regeneration provides some of the best evidence for Wolpert's model, as the cells that mediate the regrowth of

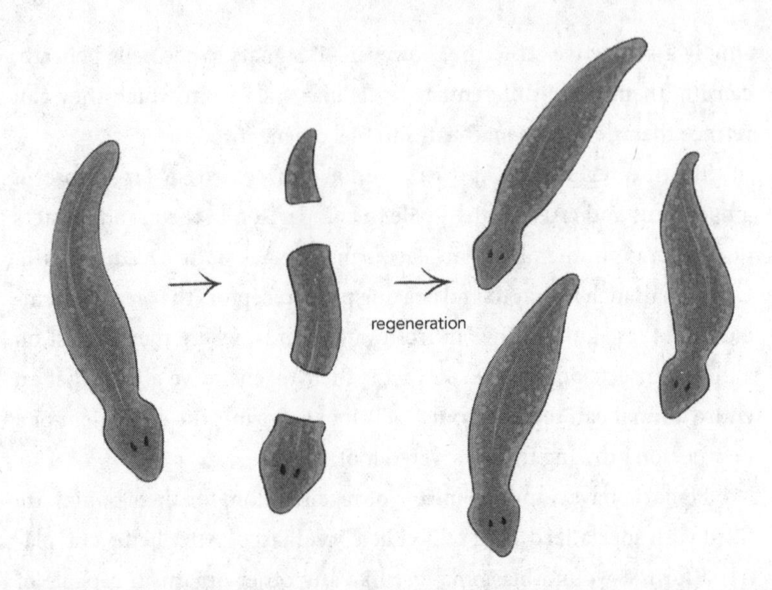

Planaria are the masters of regeneration; after a worm is cut into dozens of pieces, each piece can form a new worm.

limbs or tails behave as if they have a "positional chart"—a coordinate system with which to navigate during the reconstitution of form.

LOST IN SPACE

WHILE MAMMALS LACK THE ASTONISHING REGENERATIVE SKILLS OF our cousins, we are still capable of spectacular feats of regrowth. The liver, a remarkable organ whose activities range from metabolism to detoxification to the production of most of the proteins in our blood, offers one of the best examples of mammalian regeneration.

The case of a patient I cared for several years ago, whom I'll call Lisa M, provides an illustration. Lisa had overdosed on Tylenol, becoming one of the 50,000 individuals in the United States who each year take a potentially lethal quantity of the over-the-counter drug. On its own, Tylenol is not damaging to hepatocytes, the principal cell of the liver. Rather, it is a metabolite of the drug—a chemical analog made by hepatocytes, ironically—that exerts Tylenol's toxic effects. By the time Lisa M sought medical attention, the toxins were already in her system, leaving the medical team with nothing to offer beyond "supportive care"—fluids, electrolytes, and drugs administered in an effort to minimize the damage. We performed twice-daily measurements of her liver enzymes—proteins whose elevated levels in the blood serve as a marker of hepatocyte demise—knowing that with every increase, more and more of her liver had died. But after several days, her condition started to improve. The levels of liver enzymes in her blood started to decline, and she became more alert. After a week she had regained her energy and appetite, and after two weeks she was discharged, on her way to a full recovery. An examination of her liver a month later would have revealed scant evidence of the prior injury.

The liver regains its mass and function through compensatory growth, not epimorphic regeneration. If a portion of the liver—a "lobe"—is removed surgically, the liver returns to its preoperative size over time. But this recovery does not involve the regrowth of the resected lobe. Instead, the hepatocytes in the remaining liver lobes compensate by proliferating or increasing in size (*hypertrophy*). We still have little understanding of how the liver senses the need to regenerate, or how it "knows" to stop regrowing once it has achieved its normal size; these gaps in knowledge reflect our general ignorance when it comes to the mechanisms that control the size of organs and organisms, as we explored earlier.

The Greek Titan Prometheus is something of a patron deity for scientists who work on regeneration. According to myth, Prometheus stole fire from the gods and shared it with humanity. For his audacity, Zeus ordered him tied to a rock whereupon a large bird—an eagle—picked away at his liver each day. As Prometheus's liver regenerated every night, Zeus's punishment became a perpetual cycle that repeated for eternity. The story is quite remarkable for both its specificity and its accuracy, for the regenerative properties of the liver far exceed those of any other solid organ. Indeed, experiments involving serial transplantation of hepatocytes—the major functional cell type of the liver—suggest that a single mammalian liver has the potential to give rise to more than a million new livers.

But therein lies a paradox. If the liver is so good at regenerating, then why do livers ever fail? Why is there ever a need for liver transplantation? The answer comes down to the nature of the harm the liver has suffered, and whether it is acute or chronic. When subjected to acute damage, even an injury as grievous as Lisa M's Tylenol overdose, a liver that was normal beforehand can engage its innate regenerative powers with full force. But chronic injury—the result of prolonged alcohol use, persistent viral infections, or other diseases—creates an environment that renders the organ's regenerative efforts futile, a consequence of the same unwelcome partner of chronic injury that impedes sprouting in the spinal cord: scarring.

USING COMPENSATORY GROWTH, MAMMALS CAN REGENERATE THEIR organs to some degree, the liver being the most extreme case. But mice, elephants, and human beings cannot regrow whole limbs as newts, salamanders, and worms can. Why might this be so?

Some researchers hypothesize that nature preserved epimorphic regeneration only in those species for whom it provided a selective advantage. This makes sense for crustaceans and amphibians, where the ability to jettison and replace a trapped appendage—a claw or a tail—might aid in the survival of the species. And perhaps it also explains why the mammalian liver possesses such exceptional regenerative abilities. Omnivorous land animals ingest a variety of foreign substances—so-called xenobiotics—whose chemical by-products have the potential to do great harm. As the liver is the principal site where such chemicals are detoxified, it is especially prone to injury. Given such a constant assault from noxious compounds, it is perhaps no wonder that nature rewarded the liver with impressive regenerative powers.

There are other hypotheses to explain why mammals may have "lost" the ability to perform epimorphic regeneration. Some invoke the transition from cold-blooded to warm-blooded animals, which occurred more than 200 million years ago, or the emergence of a more complex immune system even earlier, where the loss of regenerative capacity may have been a trade-off allowing these new biological abilities to be acquired. It has also been suggested that the ability to regenerate disappeared as animals grew larger and needed more robust mechanisms to reduce the risk of cancer. Yet none of these theories fully explains why our abilities are so limited compared to our more regenerative ancestors.

In our earlier discussion of morphogenesis, we saw that embryonic cells enjoy a knowledge of position, a vestibular sense with which they perceive where they are and where they need to be. Without such positional identities, embryonic cells would expand and differentiate but would never shape themselves into a recognizable or functional form. A teratoma is an example of what happens if a growing developmental system surrenders its notion of space.

Our evolutionary cousins that are capable of epimorphic regeneration have an ability to safely stow away positional information throughout life, and we are now beginning to understand the molecular nature of these internal compasses. One of the most interesting insights to emerge is that the nerves that run inside the salamander limb are a warehouse for these spatial memories. Within the nerves, the expression of several proteins varies along the length of the limb—highest at the shoulder and lowest by the digits—where they provide a kind of longitudinal code. By measuring the concentration of these proteins, a cell within the limb can, in effect, "know" its position relative to trunk and fingertips. There are likely to be many other factors that contribute to such topographical maps—protein gradients, electrical currents, or genomic imprints that provide cells with positional cues—which will undoubtedly be uncovered as researchers continue to study amphibian regeneration.

The prospects for human regeneration are not as promising. There is scant evidence that adult human tissues retain the three-dimensional maps of body and space that guided them during development. And if such maps—so crucial during embryogenesis—are forgotten once development ends, then no amount of cajoling would give our cells the directions they need. Our inability to regrow arms, legs, and other body parts may, in the end, be the simple consequence of faulty memory.

NEW ORGANS FROM SCRATCH?

MORE THAN 30 YEARS AGO, SHORTLY AFTER I STARTED MEDICAL school, I was having a drink with my friend Rick. We were former classmates and, being geeky fans of science fiction and technology, we liked to speculate about the future. Our conversations touched on subjects ranging from the human genome to information technology, and imagination rather than facts dominated the exchanges. This liberated us, for without having to worry about precedent or practicality, we could be as outrageous as we wanted. Most of the ideas we hit upon were nutty.

One evening, Rick was giving me a hard time about what he considered the primitive state of medicine. Specifically, he couldn't understand why humans don't routinely live to be older than a hundred. His reasoning was as follows: if most of us die because our organs fail (which is more or less true), we should be able to prolong life by replacing failing organs with new ones. When your car breaks down, you bring it into the shop, where a mechanic replaces the nonfunctioning part. Why should it be any different for the human body?

It was, and is, a logical argument. Organ failure represents the final common pathway for disease, and it is why we eventually succumb to illness. Organs fail for many reasons—infections, blood clots, blocked blood flow, immune destruction, wear and tear, infiltrating tumors, and toxicity. Physicians can mitigate the effects of end-organ damage with drugs, but when it comes to the long list of degenerative diseases caused by chronic loss of function—atherosclerosis, diabetes, Alzheimer's disease, Parkinson's disease, renal failure, macular degeneration, muscular dystrophy, emphysema—medicine's recourse is to "manage" the problem rather than to cure it.

Tissue replacement is a cure.

Consider, for a moment, patients with end-stage liver disease, who are often comatose from the toxins they cannot metabolize and yellow from the bile they cannot excrete. It is not hard to recognize those who have crossed a "point of no return" and are unlikely to survive their hospital admission. But for a subset of patients in this state—those lucky enough to undergo liver transplantation—the outcome is different. Such patients awake from their comas within days of receiving a new liver and may return home within a week or two. Most live for decades. A successful organ transplant is the medical equivalent of replacing an essential car part, with lifesaving consequences.

Transplantation is far from a panacea. The need for new organs far outstrips supply, and each year in the United States tens of thousands of individuals die waiting to receive a new liver, kidney, heart, or lung. An organ must be in near-perfect condition for it to be transplanted; consequently, only a fraction of the organs that become available for transplantation are acceptable. Conversely, many of those in need of a new organ are not suitable transplant recipients (failure of one organ is often associated with the failure of others, making the risk of the surgery itself too great). Immune compatibility and organ rejection are constant concerns for those who are lucky enough to receive transplants, requiring the lifelong use of immunosuppressive drugs. Finally, social issues come into play—a donated organ is a precious resource, and potential recipients are carefully screened to identify those who will be good "stewards" of such a lifesaving gift.

To answer Rick's question—Why can't we prolong human life by replacing burnt-out organs?—I summoned what little knowledge I had at the time. The human body is not a car, I insisted, and organ transplantation, with all its problems, is the best solution we have. And even if the technical aspects of transplantation were to improve, issues of supply and demand would always be a limiting factor.

"The human body doesn't come with a warranty, and you can't just order a new part from the dealership," I said.

But Rick was an engineer, and he smiled.

"If there aren't enough organs to go around," he said, "then why not make new ones from scratch?"

THE EARLIEST AND SIMPLEST FORM OF FABRICATED TISSUE IS THE prosthesis, a device first employed by the ancient Egyptians and Persians to replace limbs lost in battle. The past two to three decades have seen the development of sophisticated artificial hands and legs that can approximate (and in some cases outperform) the natural limb. Prostheses have come to replace damaged heart valves, blood vessels, joints, and other tissues. Cochlear implants can help deaf people perceive sound, and penile implants can reverse erectile dysfunction. If a body part has suffered a structural defect or a mechanical failure, a prosthesis can be a lifesaver.

But most organ failure is functional, not structural—the result of a breakdown in cellular activity—and in this setting, prostheses tend to fall short. Hemodialysis, the "artificial kidney," is something of an exception. Dialysis serves as a lifeline to more than a half-million Americans with kidney failure, and it provides a mechanical solution to a functional problem. But in practice, it is a half measure. The normal kidney contains over a million microscopic units called *nephrons*, each of which acts as an independent mini-organ. Nephrons filter the blood through a series of tubes, retaining some molecules (water, sodium) and eliminating others (urea, potassium) as the filtrate moves its way along the tubular network. Willem Kolff, who developed dialysis in the 1940s, realized that this function could be emulated on a larger scale by filtering the blood through a porous membrane. But despite its remarkable impact on medicine, dialysis doesn't come close to matching the function of a normal kidney, and patients who receive a kidney transplant to treat renal failure live a decade longer, on average, than those who receive dialysis. Thus, the best example of an artificial organ that we have—hemodialysis—is merely a stopgap measure, and other devices meant to replace organ function, like the artificial heart, have a far worse track record.

Certainly, we should be able to do better, right?

Standing in our way is the fact that our tissues are immeasurably more intricate than any invention, structure, or product humankind has ever made. Every nephron is a masterpiece of engineering, its twisty tubes able to secrete or absorb substances thanks to the brilliance of its three-dimensional design. This intricacy of form and function is what makes creating an artificial organ "from scratch," as my engineer friend Rick proposed, so problematic. It took evolution millions of years to finalize the plans for our internal organs, and despite the feats of technology that have set the last century apart from all that preceded it, the ability to rebuild an organ with any level of precision remains far beyond our reach. As a result, bioengineers have increasingly turned away from devices for organ replacement, looking instead to the solution that nature has already provided: cells.

AN UNEVEN START

USING ENGINEERED CELLS TO REPLACE MALFUNCTIONING TISSUES falls under the heading of "cell-based therapy," and its goal is to transform the treatment of degenerative diseases. A half century ago, the thought of fabricating replacement organs—what has become the field of regenerative medicine—belonged to the realm of science fiction. Before the 1950s, bacterial cells and phage were the biologist's bread and butter. Eukaryotic cells were more finicky. But in the latter half of the twentieth century, methods for growing human cells improved, starting with the infamous HeLa cell line derived from 31-year-old cancer patient Henrietta Lacks.

Since then, cell culture has become a routine element of most biology laboratories, and a cornucopia of cell lines are available to researchers from both nonprofit and commercial vendors. Even so, most cell lines are not normal. Relocating cells from the body to a petri dish selects for changes, some of them cancer-like, that allow the cells to grow indefinitely ("immortalization"). Moreover, the longer a cell line grows, the greater its abnormalities. Consequently, most normal cells, including those with the greatest relevance to human disease, either resist culture or are unsuitable for certain studies.

Rather than be deterred, researchers seeking to develop tissue replacements have met these challenges by launching divisions, departments, and institutes dedicated to regenerative science, often with the backing of philanthropy and/or state governments. The largest of these is the California Institute for Regenerative Medicine, a multibillion-dollar initiative from the state that spans multiple universities and has funded over a thousand projects in its 17-year history. Many of the efforts falling under the regenerative-medicine umbrella have involved stem cells—either embryonic stem cells (ESCs), induced pluripotent stem cells (iPSCs), or various kinds of adult stem cells. Getting the cells to differentiate properly has been a major challenge; as we discussed earlier, existing laboratory protocols can push cells in one direction or another, but rarely do the differentiated products bear a strong enough similarity to their normal counterparts for them to be used clinically. Even bigger hurdles include the paired challenges of getting cells to arrange themselves into the proper three-dimensional structures and then integrating them into the failing tissue so that they can function as intended.

To mitigate this issue of getting synthetic tissues to integrate properly, some scientists have considered using plasticity—the alchemical phenomenon of cellular reprogramming that gave rise to iPSC technology—to convert undesirable cells (for example, scar-forming fibroblasts) to more desirable ones (for example, hepatocytes, cardiomyocytes, neurons, and so forth). Such an approach would require substantial technological innovation, as it would likely necessitate the introduction of reprogramming factors directly into the patient's tissues. But if it worked, reprogramming cells at the site where they are needed might bypass some of the issues of three-dimensionality and integration, since the newly generated cells would already be there.

These and other approaches have been tested hundreds of times, using cells, animals, and people. To date, the results have been mixed.

In 2005, Hans Keirstead, a researcher at the University of California at Irvine, made headlines with his protocol to make human ESCs differentiate into the precursors of "oligodendrocytes," a type of neuronal protective cell, with the goal of using these cells to improve SCI outcomes. When Kierstead and his team injected the cells into rats whose spinal cords had been damaged, those that received the cells showed a statistically significant improvement in locomotion and gait compared to controls. The therapy appeared to work only if the cells were administered shortly after the injury; if months passed between the injury and cell injection, the benefit vanished.

It was a result that galvanized the SCI community, and in 2010 Geron Corporation launched a clinical trial for SCI patients—the first time an ESC derivative was given to a human being. A year later, Geron ended the trial after introducing the cells into only four patients—too few to reach any conclusions. (Geron claimed that this was a strategic decision based on corporate priorities.) Among those who received the cells, none exhibited a worsening of symptoms, but none exhibited improvement either. It was not the promising start the field had hoped for.

GLIMMERS OF HOPE

NOW, MORE THAN A DECADE LATER, THINGS ARE LOOKING BRIGHTER for cell therapy. Two vignettes—one from oncology and one from endocrinology—exemplify how cells engineered outside the body have transformed, or likely will transform, clinical medicine.

The first story began in the late 1980s, when Israeli immunologists Zelig Eshhar and Gideon Gross proposed redirecting an individual's T cells—immune cells whose principal role is to eradicate cells infected by viruses—into attacking cancer cells instead. The trick involved getting the patient's T cells to take up an engineered piece of DNA that might refocus their attention on altered structures, or "antigens," specific to the cancer cells. The gene encoded by this DNA fragment was known as a "chimeric antigen receptor," or CAR.

Scientists optimized the technology over the next two decades, and in 2010, a 65-year-old corrections officer named Bill Ludwig became the first person treated with an advanced version of the engineered cells. Diagnosed 10 years earlier, Ludwig's leukemia had become resistant to every standard and experimental therapy. At the end of his rope, and with few other options, Ludwig enrolled in a clinical trial at the University of Pennsylvania, where a team of researchers including Carl June, Bruce Levine, and David Porter had designed a CAR they hoped would get Ludwig's leukemia under control.

The treatment would involve several steps. The researchers first had to isolate T cells from Ludwig's blood, grow them in a clinical laboratory by the billions (T cells fare better in short-term culture than many other types of cells), and introduce DNA molecules containing the CAR gene. Then, the genetically altered cells would need to be infused back into Ludwig's body. Finally, the researchers would need to wait and see whether the technique had accomplished anything.

At first, there was no effect. Then, after several days, Ludwig's condition deteriorated. He was transferred to the intensive care unit, and it appeared that he would not survive the therapy. The CAR-bearing T cells had seemed to make matters worse. But then, after veering precariously close to death, his condition improved. What had initially appeared to be a sign of failure had instead been a harbinger of success. For in carrying out their intended responsibilities—killing leukemia cells by the millions—the engineered T cells had generated a massive inflammatory response, a side effect of a productive antitumor response. A month after receiving the CAR T cells, a bone marrow biopsy showed no sign of the leukemia. Both he and his physicians were in disbelief. For all intents and purposes, Ludwig was cured.

CELL THERAPY'S MORE RECENT SUCCESS STORY INVOLVES TYPE 1 DIAbetes, a condition in which the body's T cells get a false signal to attack the insulin-producing beta cells of the pancreas. After this "autoimmune" process kills more than 75 percent of the beta cells, the remaining cells can no longer keep up with the body's demand, and diabetes ensues. (Insulin lowers the levels of glucose in the blood, and so without it, sugar levels become alarmingly high.) Insulin, discovered by Frederick Banting and Charles Best in the early 1920s, transformed type 1 diabetes from a uniformly lethal disease to a chronic condition. But although it has served as a lifeline for diabetes sufferers for the past century, insulin comes with its own risks. Most serious of all, insulin injections can easily cause blood sugar to plummet to dangerously low levels ("hypoglycemia"), with coma and death as all-too-common consequences.

An alternative therapy for those with especially severe type 1 diabetes is "islet transplantation," a procedure in which "pancreatic islets" (cellular aggregates within the pancreas that contain the beta cells) from an organ donor are transplanted into the patient's liver. The relative success of this procedure proved that beta cells, obtained from an outside source, could be an effective treatment for diabetes, even if the cells were situated somewhere other than their original anatomical position in the pancreas. But transplantable islets are even harder to come by than kidneys, livers, or hearts, making supply an uncompromising limiting factor. Consequently, regenerative-medicine researchers embarked on a mission to find alternative sources of beta cells.

ESCs were the most obvious candidate. After Thomson and Gearhart's success at deriving human ESCs in 1998, multiple labs began to design and test differentiation protocols, looking for conditions that could coax the pluripotent ESCs to differentiate into insulin-producing cells. It was slow going at first, and not until a decade later did the effort gain true momentum, much of it driven by Doug Melton, a stem cell researcher at Harvard. (Melton was my postdoctoral adviser from 2000 to 2006.)

By 2015, the differentiation protocols contrived by Melton and others had advanced far enough for researchers to start thinking about evaluating them clinically. In 2021, the biotech company Vertex launched a clinical trial to test the effectiveness of their "fully differentiated, insulin-producing islet cell therapy" in diabetes. Patient number one was Brian Shelton, a 64-year-old mail carrier whose hypoglycemic episodes had become so severe that collapses and seizures had become an almost daily event. Uncertain whether each day might be his last, Shelton volunteered for the Vertex trial. Researchers instilled millions of cells into his liver, following protocols as if the infusion had contained islets from an organ donor instead of processed cells from a laboratory. Three months after his infusion, the company provided a peek at the early results. Not only had Shelton not suffered any adverse consequences, but his blood sugar was nearly normal. By nine months, Shelton no longer needed insulin.

THE FDA APPROVED CAR T TREATMENT IN 2017, THE FIRST TIME AN engineered cell therapy has received the regulatory body's stringent stamp of approval. Since Ludwig's courageous roll of the dice, more than a thousand people with leukemia or lymphoma have been treated with CAR T cells, with the vast majority experiencing clinical remission or frank cure. Cell therapy for diabetes is at a much earlier stage of development, and only time will tell whether the treatment has permanently liberated Shelton from the shackles of insulin. But in the meantime, it is hard to imagine a more promising beginning.

These are impressive strides forward for the treatment of leukemia and type 1 diabetes. But what do these advances mean for Karen Miner and the millions of people who currently suffer with spinal cord injury, dementia, and other degenerative diseases—not to mention the billions who will suffer from these conditions in the future if better treatments do not become available? Is it possible that these two examples will turn out to be exceptions—"low-hanging fruit"—and that other victories for cell-based therapy will be even harder fought?

It's hard to know. There have been more than 40 clinical studies using stem cells to treat cord injury since Geron's aborted trial, and although injecting cells into the spinal cord appears to be safe, no front-runner therapies have emerged. Moreover, unlike patients with leukemia or type 1 diabetes, patients with SCI often recover some degree of function over time, making it difficult to differentiate between clinical responses from a treatment and spontaneous improvements.

There remain several barriers to success. Safety is one concern. You will recall that ESCs, injected into various parts of the body, can form teratomas. Consequently, clinical researchers using stem cell derivatives for therapy will need to make sure that their cellular products lack this tumor-forming potential. And while secondary malignancies associated with CAR T-cell therapy appear to be rare, they remain a theoretical possibility. Regulators will certainly be weighing these risks carefully as they appraise this new world of cell-based therapy.

The more significant hurdle, in my opinion, is the fact that a cell's position in space can be just as important as its standing in the cellular society. Positional identity may not matter for CAR T cells (which find their way around the body naturally) or for stem cell–derived therapies for diabetes (as insulin-producing cells appear to function perfectly well in the liver). But in most cases, tissues and organs must possess a precise three-dimensional structure for them to function properly. And here, our technical abilities fall short.

Once again, this brings us back to normal development. If the biggest long-term challenge to regenerative medicine is three-dimensional space, then the embryo—with its ability to encode positional information—holds the solution. The task that awaits us is to decrypt these spatial codes so that we may (re)fashion body parts at will the way our amphibious cousins do.

For her part, Miner doesn't really care where the breakthrough comes from, just that it happens eventually. "I am still one of the most fortunate people on the earth," she reflects, a note of incredible optimism for one who has suffered so much, and she remains hopeful that better treatments for individuals with spinal cord damage are coming—if not in her lifetime, then for the benefit of future generations.

"The body knows how to heal itself," she concludes. "It just needs science to help it along."

Chapter 11

DAY SCIENCE AND NIGHT SCIENCE

If you hear a voice within you say "you cannot paint," then by all means paint and that voice will be silenced.

—VINCENT VAN GOGH

Discovery can be a messy business. Rarely is there a straight line to new knowledge; instead, hypotheses and interpretations must navigate a roundabout and often directionless path. But rather than an obstacle to wisdom, this is better considered a necessary part of the process. We have already seen several instances in which important discoveries came about when researchers abandoned their original ideas, creating bias-free conditions that fostered new paradigms. Cells, genes, regulation, induction, transcription, stem cells, reprogramming—all were amorphous inklings, closer to guesses than theories before they became bedrocks of biology.

Researchers and writers have proffered various metaphors to describe this ethereal quality of science, but my favorite comes from François Jacob. In his memoir, *The Statue Within*, Jacob divides scientific inquiry into two categories, which he refers to as "day science" and "night science":

> Day science employs reasoning that meshes like gears and achieves results with the force of certainty. One admires its majestic arrangement as that of a da Vinci painting or a Bach fugue. Conscious of its progress, proud of its past, sure of its future, day science advances in light and glory. Night science, on the other hand, wanders blindly. It hesitates, stumbles, falls back, sweats, wakes with a start. Doubting everything, it feels its way, questions itself, constantly pulls itself together. It is a sort of workshop of the possible, where are elaborated what will become the building materials of science. Where hypotheses take the form of vague presentiments, of hazy sensations. It is impossible to predict whether night science will ever pass to the day condition. When that happens, it happens fortuitously, like a freak. What guides the mind then, is not logic. It is instinct, intuition.

Uri Alon, a biologist at Israel's Weizmann Institute, has another name for the space where basic assumptions break down and new concepts can be born—a realm he calls "the cloud." Scientists who find themselves in the cloud are filled with anxiety, like hikers lost in the woods craving the safety of the trail. It is an untethered feeling, the sense of being in a room with no exit. It is especially unsettling for students, whose preconceptions of laboratory research are grounded in a cycle of hypothesis, experiment, and refined hypothesis. If only it were that simple! Instead, nature reveals its most closely guarded secrets preferentially to those who can sit with uncertainty, while more rushed visitors are likely to miss out.

The importance of meandering as a prerequisite for discovery becomes even more evident if we compare basic science, the type of exploratory research we have mostly focused on, to engineering, with its defined goals and timelines. Bill Kaelin, winner of the 2019 Nobel Prize in Medicine, described the difference succinctly, noting that engineers are accustomed to "using the rules" while basic scientists are more interested in "learning the rules." At its most fundamental level, engaging in basic science is an effort to learn a new language—one whose very existence is unknown at the outset.

Of course, this type of fundamental research is not without application. The most profound advances in medicine over the past 20–30 years can trace their origins to studies in plants, bacteria, phage, and flies—work that an uninformed observer at the time might have viewed as irrelevant. We need look no further than the COVID-19 vaccines—mRNA preparations that owe their existence to a chain of discovery starting with Jacob's phage work—to grasp the power of night science.

These distinctions have implications for the biomedical research enterprise, where the thirst for impact, or "deliverables," draws resources from more open-ended projects that may carry greater long-term (albeit less visible short-term) potential. "Moonshot" operations—vast initiatives involving dozens or hundreds of scientists with a single-minded goal—are based on the engineering premise that money and sweat can fill a knowledge gap. To be clear, there is a place for such undertakings, especially when the goal is well defined. This was the case for the Human Genome Project—the multibillion-dollar venture to determine the complete DNA sequence of human beings. But when the path of discovery is less clear—how to cure cancer or autoimmune disease or how to comprehend human consciousness—night science, despite its slow pace, is where breakthroughs come from.

"Basic scientists are increasingly asked to certify what they would be doing with their third, fourth and fifth years of funding, as though the outcomes of their experiments were already knowable," Kaelin writes. It is a kind of thinking that neglects the unpredictability of new knowledge and ignores the history of night science as the disruptive force behind discovery.

THE INTRICACIES OF CELLULAR MEMORY

THE GREAT PARADOX OF NIGHT SCIENCE IS THAT IT IS INVISIBLE before its transition to daytime. The closest we can come, prospectively, is to consider a problem that lives at dawn, neither shielded by darkness nor fully illuminated by sunlight. The example I have in mind is *epigenetics*, a rapidly growing field that can best be understood in the context of two conflicting facts we already encountered.

Fact number one: All cells contain an almost identical complement of DNA. This is the principle of genomic equivalence, which ensures that a cell retains a complete set of genes as it differentiates.

Fact number two: Cells have *memory*. As the embryo grows and its cells differentiate, the citizens of our cellular societies hold fast to their occupations and pass them along to their children.

A consequence of fact number one is that cells carry *excess* information. Hepatocytes possess all the genes needed for neuronal function, and leukocytes have all the information needed to produce cartilage. This creates peril for fact number two, for it leaves open the possibility that at any moment a cell might go "off script" and express an inappropriate cellular program. Imagine the bodily chaos that would ensue if our cells started to forget who they were—if liver became lung or brain turned to bone. In principle, such dramatic transformations are possible, as Gurdon and Yamanaka showed us with animal cloning and cellular reprogramming. But under normal circumstances, this does not occur. Cells hold fast to the identities conferred during development and pass those identities along to their progeny.

Cellular phenotype is a function of gene expression—which genes are ON, and which are OFF. Gene expression, in turn, is regulated by DNA-binding proteins, the transcriptional activators or repressors we learned about in Chapter 4 that regulate the production of mRNA. For bacteria and phage, these gene regulators were all that was needed for a cell to adapt to its environment. When Jacques Monod switched the diet of *E. coli* from glucose to lactose and back again, the organisms happily flipped back and forth; transcription factors alone mediated the response to changing food sources. The microbes had little need for memory.

But in multicellular organisms, where occupational recall is essential, cells had to find a way to remember their identities and bequeath the information to their descendants. The ability to pass differences in gene expression to the next generation, all in the face of a uniform genome, was yet another facet of the One Cell Problem for nature to solve. It did so, quite ingeniously, with a phenomenon known as epigenetics.

In the early 1940s, biologist Conrad Waddington coined the term "epigenetics" to describe a different mode of heredity, one in which information could be passed down between cells in the developing embryo but not between animal generations. Waddington's proposal preceded the recognition that DNA is the genetic material, and so his ideas lacked molecular footing. But he perceived that developing cells must have a way to transmit information as they divided—information regarding which genes should be ON and which should be OFF. Epigenetics encapsulated the notion that cellular memory is maintained through a new mechanism of heredity, one distinct from the genetics of Mendel, de Vries, and

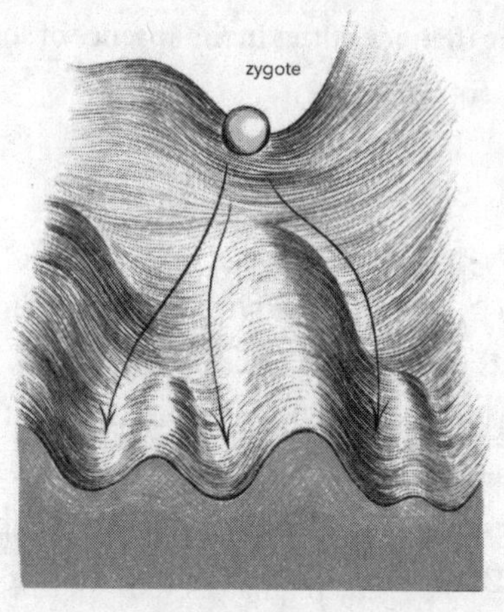

Conrad Waddington suggested using the imagery of a landscape for considering how cell fates are acquired during development. In such an "epigenetic landscape," the zygote starts out at the top of a hill. As that cell divides, its descendants make their way down the slope, deviating right or left based on signals received from neighboring cells. Cells that come to rest in a distant valley are the specialized cells of the cellular society–retina, cardiomyocyte, or neuron, for example. Because substantial energy would be required to move a cell from one valley to another, adult cells exhibit far less plasticity than embryonic cells.

Morgan. In other words, while genetic mechanisms mediated the heritability of animal form during evolution, epigenetic mechanisms mediated the heritability of cellular identity during development.

Waddington provided a visual aid to help biologists understand this concept—a metaphorical "epigenetic landscape" that imagined cells as balls rolling downhill as they differentiated. In the standard "Waddington diagram," balls at the top of the hill are meant to represent multipotent cells with the potential to generate many types of offspring, while the valleys below represent those different fates. As cells start their downward journey, inductive forces—signals received from neighboring cells—nudge them in one direction or another. A cell that has access to many paths at first (plasticity) would find its ability to end up in other parts of the landscape growing smaller and smaller as it rolled downhill (commitment). Cellular memory, according to Waddington, is a kind of molecular gravity—it prevents cells, and their progeny, from changing their identities in the absence of some extraordinary force of displacement.

THE REVOLUTION IN MOLECULAR BIOLOGY OF THE 1950S AND '60S SET the stage for molecular studies of cellular memory, starting with the proposition that a chemical modification of DNA, rather than a change in the sequence itself, might be responsible. An early candidate was *DNA methylation*, a chemical reaction that results in a slight alteration of the cytosine base of DNA. During DNA methylation, a carbon and three hydrogens (a "methyl group") are added to cytosine, converting it to a chemical analog known as 5-methylcytosine. Researchers found that 5-methylcytosine is distributed unequally across the genome, abundant near some genes and scarce near others. Consequently, they speculated that cells might use the presence or absence of 5-methylcytosine as a marker—a molecular asterisk—to distinguish which genes to express and which to muzzle.

In the 1970s and '80s, Adrian Bird, Howard Cedar, Aharon Razin, and others built a strong case for the model. They compared genes with a wealth of 5-methylcytosine (that is, genes in which many of the cytosines within or surrounding the gene were methylated) to those with a paucity of 5-methylcytosine. Genes with a low level of methylation were expressed at high levels, while those with a high degree of methylation were expressed at low levels, suggesting that DNA methylation inhibits gene transcription. Even more striking, they found that a given pattern of DNA methylation was passed from one cell generation to the next; genes that were highly methylated remained so through repeated rounds of cell division. This chemical mark is thus one repository of cellular memory, a simple chemical tag by which cells heritably establish which genes should be ON (by leaving them unmethylated) and which should be OFF (by rendering them highly methylated).

Methylation is a major mechanism of epigenetic recall, but for it to operate correctly, cells must begin with a clean slate. It turns out that early in embryonic life, mammalian embryos "erase" almost all methylation marks inherited from the father's sperm and the mother's egg. This gives the embryonic cells a fresh start from which to establish their own methylation patterns during differentiation. All animal cells contain enzymes that can add, remove, or maintain methyl groups from cytosine, and their activity is essential for cellular identity. Yet despite all we have learned about the mechanics of cellular memory, an important question remains unanswered: How do cells determine which genes to methylate (and therefore silence) and which genes to leave unmodified (leaving them capable of being expressed)? In other words, how do cells acquire their memories in the first place?

REIMAGINING INHERITANCE

IN PARALLEL WITH THE REVELATIONS ABOUT DNA METHYLATION IN the final decades of the twentieth century, a more intricate, and likely more ancient, information-storage method was being discovered. But instead of engraving reminiscences into DNA, this memory system involved the proteins with which the genetic material is intimately bound—proteins called *histones*. Unlike *prokaryotic* bacteria, eukaryotic cells do not allow their DNA to float freely. Rather, DNA associates with histone proteins to form specialized structures within the cell nucleus. These subunits, called *nucleosomes*, consist of repeating bead-on-a-string-like subunits in which the DNA wraps itself around histone cores every 200 or so nucleotides. This complex of protein and DNA is known as *chromatin*.

Nature devised this seemingly complicated arrangement to deal with a storage problem. As anyone who has tried to fit weeks' worth of clothes into a carry-on bag can tell you, packing strategies matter. There are

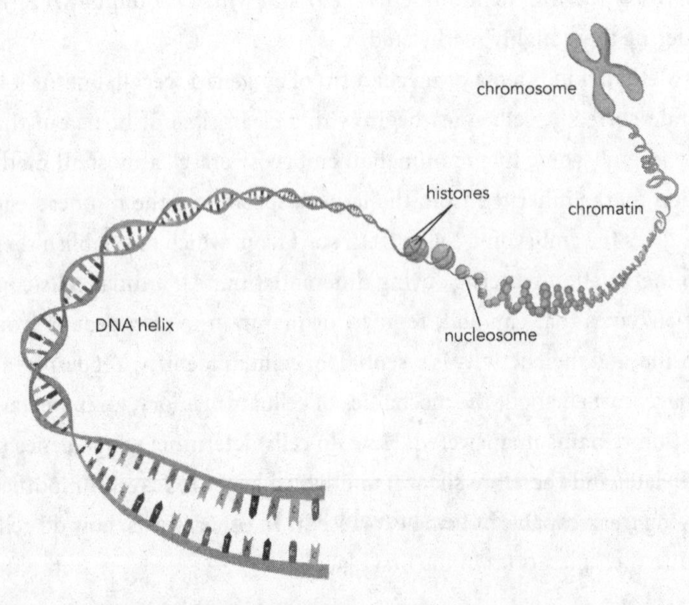

The sheer amount of DNA present in every cell means that it must be efficiently packaged. Within the nucleus, DNA is not free to move around but instead wraps around a protein complex consisting of eight histone proteins; each repeating unit is called a nucleosome. Chemical modifications to histone proteins influence the extent to which a gene is transcribed into mRNA—an epigenetic mechanism distinct from DNA methylation, which allows a cell to maintain and pass on its identity.

various methods to maximize space (personally, I prefer the "ranger roll" method). But the worst method is no method at all—randomly stuffing belongings into a suitcase. In humans, the six billion nucleotides that make up the genome of a single cell would extend for six feet if stretched out as one linear molecule. Without an efficient packing strategy, this amount of DNA would never fit inside the nucleus; nucleosomes provide an indispensable organizational scaffold.

But nucleosomes endow cells with another benefit—an additional source of molecular memory. In the 1960s, researchers obtained evidence that, like cytosine bases in DNA, histone proteins could also be chemically modified. Methylation is one such modification, but others were found. For example, histone "acetylation" results in the addition of an acetyl group, a chemical tag slightly larger than a methyl group. Each modification changed the structure of its corresponding nucleosome in a subtle but unique way.

At first, these modified histone proteins were mere curiosities. But in the 1990s, studies from an unexpected source—the brewer's yeast, *S. cerevisiae*—proved their importance. The first piece of evidence came from geneticist Michael Grunstein, who showed that a yeast cell's ability to transcribe a gene was directly related to its ability to acetylate histone proteins. Then, in 1996, researchers David Allis and Stuart Schreiber independently found that enzymes in yeast that add or remove acetyl groups from histones have mammalian counterparts, with similar effects on gene expression. Once it became clear that the pattern of histone modifications across the genome could be inherited from one generation to the next, just like 5-methylcytosine, histones joined DNA methylation as a mechanism of cellular recall.

The field has exploded since then. In addition to acetylation and methylation, other modifications to histones have been found—the addition of phosphorus (phosphorylation) or a small protein called ubiquitin (ubiquitylation). Unlike DNA methylation, which occurs only on cytosine bases of DNA, histone modification can occur on any of several amino acids within each histone protein, and each version has a different effect on gene expression. All told, there are well over a hundred unique types of histone modification. Each change, on its own, has only a modest effect on the transcription of a gene into mRNA. But added together, the summation of histone alterations can turn a gene OFF or ON.

The biologist Nessa Carey has provided the best analogy I have seen for visualizing histone biology and its effects on gene expression. Imagine a chromosome as the trunk of a large Christmas tree, she writes, with its branches sticking out as extensions of the histone proteins and its lights representing individual genes. If we then decorate the tree with ornaments, so that each branch has a different array of colored balls, stars, snowflakes, and tinsel (histone marks), each of the light bulbs (genes) would find itself in the vicinity of a different assortment of ornaments. Now, imagine that the brightness of each light is determined by that combination of surrounding ornaments, which act collectively as a dimmer to make the bulb brighter or fainter. If we now think back to the genome, considering the brightness of each light to reflect the expression of each gene, we can see how different histone decorations—what Allis nicknamed the "histone code"—give cells another epigenetic mechanism for storing information.

The regulatory complexity embedded within the histone code is dazzling. The number of permutations possible from different combinations of more than 100 histone marks is vast (at least 100^{100}). At present, scientists have a firm handle on the activity of less than 10 percent of all histone modifications, and understanding what each mark does—alone or in combination with other marks—occupies the undivided attention of dozens of laboratories. It will be years, if not decades, before we understand the histone code with the same clarity that we understand the genetic code. Nevertheless, the ultimate answer to the One Cell Problem lies buried deep within the complexities of epigenetic regulation, and with it the ability to redirect cellular destinies.

AFTER THAT RATHER TECHNICAL DIVE INTO EPIGENETICS, YOU MAY be wondering about the practical implications of DNA methylation and histone modifications. I offer a three-part answer. First, it is important to remember that gene regulation is the defining feature of cellular identity. A cell's place in the cellular society is largely a function of the genes it expresses, just as the work of a doctor, lawyer, or carpenter is, to a great extent, defined by the skill set they possess. We still do not understand how the embryo decides which genes in which cells should receive which epigenetic marks (such decisions almost certainly involve an interplay of nature and nurture, the cell-intrinsic and cell-extrinsic cues that developing cells receive). Nevertheless, we know that the *epigenome*—the summation of epigenetic marks across the genome of a given cell—determines both fate and function, acting collaboratively with transcription factors to determine, for each gene, whether it should be expressed.

Epigenetics also provides the potential to exert control over cellular phenotypes. Since cellular function depends on gene expression, the ability to selectively modulate transcription could have great medical benefits. Imagine a drug that could silence expression of a tumor-promoting oncogene in a cancer patient, or one that could activate the expression of genes encoding wound-healing growth factors in a burn patient. As we learned earlier, transcriptional regulators (activators and repressors) govern gene expression, but it has proved difficult to develop drugs that can alter the activities of these DNA-binding proteins. Drugs work by binding to specific molecular targets in a cell, a specificity that comes from the drug's ability to associate with its target(s) at critical docking sites, or "binding pockets" (akin to a key "finding" its lock). The chemical structures of most transcription factors are such that they lack exploitable binding pockets; consequently, control of transcription has historically been a no-man's-land of drug development. Enzymes, by contrast, make excellent drug targets. Enzymes are designed to catalyze chemical reactions, which they accomplish by binding their substrates with great specificity. This means that every enzyme, by definition, contains a binding pocket. Therein lies the great opportunity for the field of epigenetics, as the molecular mediators of epigenetic regulation—whether DNA methylation, histone acetylation, or other modifications—are all enzymes. Gene expression, previously outside the reach of medical drug development, is now fair game.

Finally, epigenetics is causing us to rethink how traits are passed from parents to children. As Darwin was rallying support for his theory of natural selection, one of the headwinds he faced was Lamarckism—Jean-Baptiste Lamarck's eighteenth-century model postulating that environmental forces have a direct impact on evolutionary trajectories. Of course, natural selection won out, and it does a magnificent job of explaining how new species come into existence. But studies over the past few decades have made us reconsider a once-heretical idea: the notion that traits acquired during an animal's lifetime can be inherited by future generations. This phenomenon is known as *transgenerational inheritance*, and supporting evidence comes from multiple organisms. In plants, for example, DNA methylation patterns can be inherited across hundreds of generations (contrasting with the erasure that happens early in mammalian development).

One of the best examples of transgenerational epigenetic inheritance in animals involves mouse coat color, the trait that Mendel first studied before turning his attention to peas. (It is perhaps fitting that the trait Mendel abandoned would end up as a clear case of non-Mendelian inheritance.) In 1999, Emma Whitelaw, a geneticist at the University of Sydney in Australia, showed that an inbred strain of yellow-haired mice would, on occasion, give birth to mice with brown hair. (Recall that in an inbred strain of mice, all animals are genetically identical.) The brown-haired mice, in turn, also produced offspring with either yellow or brown hair. On face value, it might have appeared that a new mutation had occurred, one that led to either yellow or brown hair based on the alleles inherited by the offspring (according to Mendel's laws). But as she looked more carefully at the patterns of inheritance, Whitelaw saw that subsequent generations did not follow the Mendelian explanation. Instead, a mouse's probability of having yellow or brown hair turned out to depend on the mother's hair color—brown-haired mothers (and grandmothers) were significantly more likely to have brown-haired offspring than were yellow-haired mothers. In other words, the tendency to inherit one or another trait—even in mice with identical genotypes—depended on the phenotype of the mother. As Whitelaw and others investigated the phenomenon more fully, the culprit turned out to be DNA methylation—a pattern of methylated cytosines that managed to cross the generational divide.

Transgenerational inheritance is a bit harder to study in humans, but the best evidence comes from those who survived the Dutch famine of 1944–45, a period of cruel deprivation resulting from Nazi food blockades during the last winter of the Second World War. Robust record-keeping by the Netherlands health service has allowed epidemiologists to study what happened to children conceived before, during, or after the famine. The results are fascinating. Those who (as fetuses) were exposed to famine during the first trimester had a higher risk of metabolic conditions, including obesity and heart disease, compared to those not exposed to the famine or whose exposure came later in development. Of course, this is not in itself an example of transgenerational inheritance; instead, it merely suggests that nutritional deprivation early in embryogenesis has metabolic consequences that persist into adulthood. The more striking finding comes from an analysis of the next generation: children born between 1960 and 1985 whose *mothers* were fetuses during the Dutch famine. Remarkably, the firstborn babies of these women weighed more if their mothers had been exposed to nutrient deprivation during the first trimester, but not the third trimester. In other words, a baby's birth weight appears to be correlated with the intrauterine experience of its mother, when she herself was an embryo.

Epigenetics is very much a work in progress. It is fair to say that our understanding of the histone code is at a kindergarten level; we know the basic alphabet and we can read simple words. We have identified many of the enzymes with which the cell writes, erases, and reads its methyl and acetyl imprints, and we can modulate some of these enzymes' activities. But what of the future? Will we one day routinely prescribe "epigenetic drugs" that can rejuvenate a cell or redirect its behavior, thereby revolutionizing clinical medicine? Will transgenerational inheritance expose more chinks in the previously impenetrable armor of Mendelian genetics and Darwinian evolution?

It is hard to say, which is fitting, since that is the very essence of night science.

ENGINEERING HUMANITY

ONE OF THE GREAT DIFFICULTIES DEVELOPMENTAL BIOLOGISTS FACE is that so much activity is hidden from view. Cells in the embryo's interior are blocked by those lining its exterior, and molecular events are too small for us to view them directly. The problem is compounded in mammals, where the spectacle of development takes place embedded in the mother's uterus. Classical studies of development circumvented this problem by relying on embryos that develop out in the open—sea urchins, frogs, flies, and worms. Consequently, a great deal of what we know (or think we know) about human development has been inferred from studies in these and other "model organisms." But animal models have their limits. If we are to develop new reproductive technologies or find better ways to detect and prevent miscarriage and birth defects, we must understand the development of our own species with equal depth.

Studying embryonic development in humans is fraught with challenges. Ethical concerns have occupied the spotlight, as there is near-universal agreement that it is morally unacceptable to experiment on human embryos beyond a certain point in their development (although there is little consensus on when, precisely, that point is). International standards for human embryo research, established in the wake of IVF technology in the late 1970s, introduced the "14-day rule," which prohibited scientists from culturing embryos that are older than two weeks postfertilization.

There are also numerous logistical and technical challenges to human embryo research. Almost all human embryos used in research are those donated by couples undergoing fertility treatments, which makes the supply chain unpredictable. But what if there were another source—one that mitigated these ethical and supply issues? What if it were possible to study human embryonic development without using a human embryo?

Recently, technical advances in cell culture and differentiation methods have made such a scenario possible. Building on the fact that induced pluripotent stem cells (iPSCs) can adopt a variety of cell fates, several labs have developed methods for generating all the cell types present in a week-old human embryo—the epiblast, hypoblast, and trophectoderm components of a blastocyst. When cultured in three-dimensional gels, these cells autonomously self-assemble into "blastoids," rounded structures containing roughly a hundred cells. Remarkably, despite being the products of cell lines rather than zygotes, blastoids bear an uncanny resemblance to their human embryonic counterparts, including the ability to latch on to the endometrial cells that line the uterus.

Which begs the question: If it looks like an embryo, and acts like an embryo, but came from a patch of skin or a drop of blood, then what is it?

TEN YEARS AGO, A TSUNAMI SWEPT ACROSS THE FIELD OF BIOMEDICAL research. The Clustered Regularly Interspaced Short Palindromic Repeats system, known as CRISPR, was originally identified in the 1990s as a bacterial defense system against invading phage. But in 2011, two scientists—Jennifer Doudna at the University of California, Berkeley, and Emmanuelle Charpentier at the University of Sweden, Umeå—envisioned a way of repurposing this ancient microbial defense system to "edit" DNA sequences in living mammalian cells. Although less than a decade old, CRISPR has since become the go-to technique for altering genomes. It is powerful, it is fast, and it is easy.

CRISPR has already overturned some of our most-used methods for manipulating cells and animals. It previously took more than a year to create a strain of "knockout mice," using the techniques Mario Capecchi and Oliver Smithies devised for generating animal models of human disease. But now, with CRISPR, the same feat can be accomplished in a matter of weeks. Genetic screening, the broad-stroke approach that Eric Wieschaus, Christiane Nüsslein-Volhard, and Sydney Brenner used to outline the molecular basis of development, has also been supplanted by the technique, where CRISPR "libraries," instead of mutagens, are giving researchers faster and more in-depth access to important genes.

Most recently, CRISPR has made its way into the clinic, with two blood disorders—sickle cell disease and beta thalassemia—leading the way. In both diseases, mutations in the beta-globin gene cause the oxygen-carrying hemoglobin molecule to malfunction. In sickle cell anemia, these mutations cause red blood cells to deform and aggregate within blood vessels, leading to recurring episodes of severe and life-threatening pain; in thalassemia, different mutations in the beta-globin gene cause anemia.

Several research groups are now using CRISPR to correct the gene defect in hematopoietic stem cells, which can then be returned to the patient's bone marrow. This, it is hoped, will result in a new blood system derived from the CRISPR-edited cells, which could reduce the disease's severity or eliminate it altogether. And in at least one study, the results have been dramatic—a complete freedom from the agony of sickle cell "crises," the almost unbearable flare-ups of the disease. CRISPR-based approaches to other diseases are following close behind, including hemophilia, cancer, and congenital blindness.

Another frontier in which the magic of CRISPR is exerting its influence is the field of *xenotransplantation*—the use of animal organs for transplantation. As we saw in the last chapter, replacement organs for those with failing livers, kidneys, hearts, and pancreatic islets are in short supply. The result has been a cottage industry of cell-replacement and cell-plasticity approaches to treat disease. Scientists have long considered using animal organs to meet the demand, with pigs topping the list based on the comparable size of their organs and the precedent of replacing failing human heart valves with porcine ones. But several barriers have stood in the way, most notably the recognition and rejection of the transplanted animal tissue by the recipient's immune system. Several companies and their academic partners are attempting to surmount these barriers by creating genetically engineered strains of pigs whose organs are less likely to trigger an immune attack in a transplant recipient.

In 2020, clinicians at the University of Alabama at Birmingham, working with the Virginia biotech company Revivicor, implanted two kidneys from a genetically engineered strain of pigs into Jim Parsons, a 57-year-old man who was declared brain dead after a dirt-bike accident. It was a short-term experiment, but the transplanted kidneys appeared to function normally for the three days of the trial. In 2022, a more ambitious effort was undertaken when cardiac patient David Bennett Sr., also 57, received a genetically engineered pig heart at the University of Maryland Medical Center in Baltimore. (Bennett had been deemed ineligible for a conventional heart transplant.) The xenotransplanted heart functioned well, and Bennett lived for another two months. But a post-mortem evaluation revealed traces of a pig virus—porcine cytomegalovirus—in his blood. Whether or not the reemergence of this latent pig virus contributed to Bennett's death, there is clearly much more work to do before xenotransplantation can solve the transplantable-organ supply problem.

THE ADVANCES IN EMBRYO CULTURE AND GENE EDITING DESCRIBED above are a mere sampling of things to come, and as with all new technologies, they raise questions about how far the technologies can or should go. In the arena of embryo culture, there is nothing magical about the

14-day mark beyond which ethicists have proclaimed human embryo research should stop, and it is almost certain that existing or modified conditions could support embryonic growth well beyond this point. But how long is too long? Three weeks? Four weeks? Once nerve cells start to form? Once there is a heartbeat? And should it make a difference whether the entity in question is a proper embryo—the product of sperm and egg—or a blastoid whose only parent was a cultured fibroblast? And finally, who should be making these decisions?

These emerging technologies serve up questions we have never confronted before as a species, simply because they fell outside the sphere of imagination. One extreme example is the creation of "interspecific embryo chimeras"—embryos containing a combination of cells from two different species—that could serve as an alternative source of transplantable organs. (Unlike the pig donors described earlier, the organs derived from chimeras could in principle be composed primarily of human

Chimeras between sheep and goats can be made by combining early embryonic blastomeres from sheep and goat embryos, resulting in what is popularly known as a "geep." (REPRODUCED WITH PERMISSION FROM SPRINGER NATURE.)

cells.) Chimeras between distantly related species, such as humans and mice, are unable to advance very far in development due to molecular incompatibilities. But when embryonic cells from closely related species are mixed, the effort is far more likely to be successful, with the most famous example being the "geep"—a fertile chimera resulting from the mixture of goat and sheep blastomeres.

In 2021, a research team at the Salk Institute in La Jolla, California, succeeded in producing monkey-human chimeric embryos. Before being harvested after roughly two weeks, before the 14-day mark, the hybrid embryos matured to the blastocyst stage, and though they contained cells from both species (about 5 percent of the cells were human), the embryos seemed to develop relatively normally. Using different techniques, Chinese researchers introduced a human neuron-specific gene into rhesus monkeys, with claims that the transgenic offspring performed better on cognitive tests compared to their wild-type counterparts. (Ethical questions have been raised about the work.) These and other efforts, while not traversing any cut-and-dried ethical standards, are certainly testing them.

The biggest questions of propriety have surrounded the merger of these two burgeoning technologies—embryo culture and gene editing. In November of 2018, Dr. He Jiankui, a researcher in Shenzhen, China, announced over YouTube that he had created the first gene-edited babies in the world. In his video, Dr. He described the two babies, referred to as Lulu and Nana, as the offspring of a Chinese couple that had undergone a standard in vitro fertilization with one exception: the researchers had taken the fertilized eggs and used CRISPR to knock out a single gene—CCR5—before reimplanting the embryos into their mother's uterus. He was swiftly condemned by the international scientific community for what was widely viewed as an ethical breach. He was fired from his job, and a year later he and his colleagues were fined and sent to prison for the crime of "illegal medical practice." (He was released in 2022.)

He defended himself by explaining the reasoning behind his unsanctioned efforts. He and his team had chosen CCR5, he explained, because the protein it encodes is used by the HIV virus to infect cells. It is known that individuals with naturally occurring mutations in the CCR5 gene are immune to HIV infection. Consequently, zygotes with CRISPR-mediated mutations in the CCR5 gene should give rise to HIV-immune babies. (The other facet of He's defense was that the embryos he used came from a couple who were reluctant to have children because the father was HIV positive; according to He, they gave their consent for the procedure because it gave them hope of having disease-free children.)

There is a near-universal consensus that He's activities crossed a line. The two "CRISPR babies" Lulu and Nana, and a third baby born to a different HIV-positive couple, are presumed alive and healthy. But most details of the procedure and its outcome remain undisclosed. Most importantly, we do not know what long-term physical or mental health issues these infants may face in the future, a result of "off-target mutations"—accidental editing of genes other than CCR5, genes whose functions we don't yet know about.

But as with many technologies that were initially dismissed out of hand as strange and unnatural when they were introduced—vaccination, in vitro fertilization, recombinant DNA technology, and even handwashing were all initially viewed with skepticism or outright hostility—it may simply take time to understand how to use our newfound abilities in a manner that society finds acceptable.

The pace of discovery regarding epigenetic regulation and gene editing is moving too fast for even its practitioners to keep up with, leaving society at large with few opportunities to weigh the pros and cons. For the time being, standards are being established by the scientific community through professional societies like the International Society for Stem Cell Research and the World Health Organization. Regulatory bodies like Congress will weigh in eventually, but it is not too early to begin having thoughtful discussions. With time, we may find that there are constructive and ethically agreeable applications of these technologies short of throwing the (gene-edited) baby out with the proverbial bathwater.

INFORMATION OVERLOAD

IN 1958, THE PHYSICIST ROBERT OPPENHEIMER SPOKE TO A GROUP OF editors and journalists about knowledge. Oppenheimer contended that the essence and structure of knowledge is fundamentally different now from that which persisted from antiquity through the nineteenth century. In earlier times, being "educated" meant having a grasp on all disciplines—philosophy, history, mathematics, and the natural sciences. The twentieth century marked a tipping point in that equation, as the growth rate of knowledge outpaced the ability of even the most scholarly individual to keep up. Noting that Plato recommended mathematics as a tool for discriminating truth from falsehood and good from evil, Oppenheimer observed:

> Plato was not a creative mathematician, but students confirm that he knew the mathematics of his day, and understood it, and derived much from it. . . . Today, it is not only that our kings do not know mathematics, but our philosophers do not know mathematics and—to go a step further—our mathematicians do not know mathematics.

The challenge of absorbing and sharing science has only gotten worse since Oppenheimer's time. Gaps continue to widen between scientists and the public, and between scientists and other scientists. A hundred years ago, we knew next to nothing of DNA or mRNA and had only a superficial understanding of embryogenesis. It is impossible to know what the next century will bring, but the discoveries that await us are likely to be just as spectacular.

As with all technology, the benefits must be weighed against the risks (ethical, financial, or medical), a task that is increasingly difficult in the age of information overload. If scientists working in the various subdisciplines of biology barely have the time to keep up with developments in their own fields, it will be even harder for the broader society to digest and reflect upon this new knowledge. Yet it is critically important that we do so, for that is the only way we will reach a consensus on how these technologies will be used.

Biological literacy has never been more important.

MANY OF THE ELEMENTS DRIVING INNOVATION AND DISCOVERY HAVE remained constant over the past century. Louis Pasteur famously said that chance favors the prepared mind, emphasizing how luck rewards an openness to the unexpected. Bun McCulloch's resolve to study the splenic lumps of irradiated mice was one example of this favoritism, as was Roy Stevens's attention to the testicular tumors of his strain 129 mice. But chance also operates in a more prevalent and pedestrian manner, its hand guiding both research and researcher well before the first experiment is performed. This kind of serendipity was in operation when John Gurdon's efforts to study insects were thwarted again and again, leading him to become a history-making developmental biologist. And it was at work again when Ernest McCulloch the physician found himself paired with James Till the physicist, and when fate placed François Jacob and Jacques Monod at opposite ends of the hall in the institute bearing Pasteur's name. Serendipity is the magic powder that propels research forward, and it continues to play an outsize role in biomedical research.

Other features of success persist as well. The scientists in our narrative have tended to be young, and this remains typical of the research enterprise. The biggest discoveries are still likely to come from those at the very beginning of their careers, scientists who are not yet swayed by the assumptions, preconceptions, and presumed limitations that pervade every field. Environment also continues to play an important role, as brain trusts like the Station, the Attic, and the Fly Room continue to offer up plentiful servings of consultation, criticism, and collaboration, now in the form of university departments and research institutes. Finally, intellectual drive remains an essential research ingredient. While providence can play an important role in discovery, it rarely advances knowledge unless accompanied by sweat.

But even as these features of research have endured, other elements have changed since the early days of experimental embryology, when scientists like Weismann, Driesch, and Roux presided over solo operations. Gradually, this practice gave way to small-group research, where the work necessitated the combined efforts of pairs or trios. Women, historically excluded from the laboratory bench, have increasingly contributed to discovery, and now make up nearly 50 percent of the biological sciences workforce. (Racial and ethnic diversity is also on the increase, although representation in the lab is still far from mirroring that in the general population.) Now, in the twenty-first century, we have entered an era of "big science," where large groups, often working at multiple institutions, consume a greater and greater share of the research pie. This has resulted in massive undertakings like the Human Genome Project and the Cancer Genome Atlas, fueling discoveries that could never have been accomplished by individuals or small groups working independently.

But this redirection may come at a cost.

The scientists we met early in our narrative were largely driven by curiosity—a determination to understand how tissues and cells function. Whatever utility their work might have, this was not the force that drove them forward; rather, they were propelled by the unadorned thrill of the hunt. But as answers to biological questions became increasingly associated with commercial (and medical) potential, motives became more complex. The products of the biotechnology revolution—especially protein-based drugs that could be synthesized by cells—have made their way into the clinic, creating lucrative opportunities for pharmaceutical companies, which until then had relied entirely on chemistry. With these new developments, the urge to prioritize applied science, rather than knowledge for its own sake, became irresistible.

There is nothing surprising or improper about an emphasis on applied science. Given the choice between a project with foreseeable results and one whose outcome is uncertain, it would be logical to choose the former. And for patients or their loved ones struggling with disease, or lawmakers faced with budgetary decisions, the focus on science with tangible benefits is only natural.

But placing too great an emphasis on application has a downside. We have seen several times during our journey how the biggest discoveries—ones that have changed the course of biology and are in the process of transforming medicine—have been those with no intended application. One can easily imagine a modern-day pundit, transported back in time, ridiculing the discoveries we have recounted here: the seeming irrelevance of a white-eyed fly, an uncoordinated worm, or a tumor arising in the testicle of a mouse. But it is a fact that a large and increasing percentage of the therapies used in modern medicine can trace their origins to basic research whose outcomes and future applications were unforeseeable. By its very nature, knowledge with the greatest impact begins with a detour through night science, a meandering that defies predictability.

There is an opportunity, and indeed an imperative, to support both applied and basic science. Collaborations between labs in academia and industry have been highly successful at promoting the former. Hundreds of diagnostics, vaccines, drugs, and other products have resulted from such partnerships, demonstrating the benefits of goal-oriented research. But the next generation of innovation will require *new knowledge*—the truths whose existence we do not yet realize—and this can come about only through the imperfect and unpredictable efforts of basic scientists.

This facet of discovery is in danger, and as the cost of doing science has increased, budgets have not kept pace. Most funding for basic research in the United States comes from the National Institutes of Health, yet the basic funding instrument—the "modular R01 grant"—has not seen an increase in the level of support since 1999. Factoring in inflation, this amounts to a dramatic loss of purchasing power at a time when the opportunities for discovery have never been greater. Some of this shortfall has been filled by philanthropy, and the pharma industry continues to support basic science research at a small scale. But these resources have not been enough to fill the funding gap, and a decrease in the capacity to do basic research has been the result. While we can easily see the benefits of basic discoveries that have made their way into the world, it is much harder to measure the cost to society of discoveries that *aren't made*.

Development does not end with birth or even puberty, as the programs nature uses to build the embryo persist in adulthood, called into action to maintain tissues or repair them. The developmental biology of the future will involve a synthesis of the enduring and evolving attributes of research—the interplay of night science and day science, and the utilization of innovative tools and fresh paradigms. It will be more multidisciplinary than ever—involving bench scientists, mathematicians, bioinformaticians, and clinicians—and it will employ a variety of model organisms, divergent technologies, and large data sets. The research will continue to be directed by individuals with vision and persistence, scientists who wish for the same thing their predecessors sought—to coax from the embryo its most carefully guarded secrets.

When it comes to the body and its cells, our ignorance dwarfs our understanding. To an inquisitive mind, this knowledge gap is cause for celebration. After all, this is what is most compelling and beautiful about research—the excitement that comes from navigating the poorly charted terrain at the interface of ignorance and understanding. Our curiosity is unique in the animal kingdom—we are the only creatures that can ask the question "How did I get here?" The answer, on its own, has no tangible benefits, and yet it is one that we are compelled to ask. The quest for knowledge is our birthright.

Scientific research—generating a hypothesis and designing an experiment to test it—is immeasurably satisfying. It is a creative act, requiring a marriage of history, technique, and flexibility of mind. For the best science, indeed *all science*, begins with the simple admission "I don't know."

EPILOGUE: PARTURITION

Bob Vonderheide, a friend and colleague who studies cancer and immunology, teases me for having once remarked, "You are a developmental biologist, Bob, you just don't know it."

My lighthearted statement was a throwback to the notion that almost all fields of biomedical research—cell biology, genetics, physiology, immunology, cancer biology, neuroscience—have their roots in the study of embryonic development. Theodosius Dobzhansky, the Ukrainian-born biologist who helped unify evolutionary biology and genetics, is most famous for stating, "Nothing in biology makes sense except in the light of evolution." But he could just as easily have asserted that nothing in biology makes sense except in the light of development. Embryogenesis is as central to the body's operations as a building's architectural designs and assembly are to its flow and structural integrity.

Nearly two centuries have passed since we realized that life is built from cells, and embryos have been among our best teachers and guides ever since. Sea urchin blastomeres and fly larvae gave us a road map for understanding the mechanics of heredity, while a command of mouse embryology provided an entry point for modeling human diseases and mastering the alchemy of reprogramming. Thanks to the teachings of these diverse animal species, we have a much clearer picture of how we humans come into existence.

Still, there is much we don't know. Some of the most basic questions—

What controls an organ's shape and size? What determines life span? How does development foster consciousness?—remain confined to the land of night science. Although the answers to these questions may elude us for some time, we can use what we have already learned to develop new therapies. We have seen a sampling of these remedies—cell therapies for cancer and degenerative diseases, DNA editing to correct inborn genetic errors, novel reproductive technologies—but the biggest breakthroughs to come will probably be the ones we haven't yet conceived of.

Embryonic development has the effect of changing your perception of time. In our day-to-day (postnatal) lives, change comes slowly. The transitions our bodies make as we mature from toddlers to young adults and from young adults to old adults are subtle, recognizable only if we look for them across the span of months or years. But during development, things happen quickly. Dramatic changes can occur in the span of hours. In less than a day, a sheet of cells may roll itself into a tube, or an organ may "bud" like a seedling erupting from the ground. And yet, the embryo also embodies a greatly protracted timescale, as it reflects "designs" worked out over hundreds of millions of years. Observing an embryo mature is like viewing two sets of overlapping time-lapse images at once—one developmental and one evolutionary—intersecting in the present. It is a backstage peek at nature's grandest production.

Despite working with embryos for more than 20 years, I still have no answer to the question that stumped Aristotle and drove Hans Driesch to the paranormal: What life force, or entelechy, compels the inanimate substances surrounding us to self-assemble into cells, embryos, tissues, and bodies? At an intellectual level, it is easy enough to view it all as a matter of chemistry, an assortment of organic reactions, each governed by a thermodynamic calculation taking the reactants to a lower energy state. But at a gut level, this remains unsatisfactory. It is something the mind can know but not comprehend, like the near impossibility of perceiving that there are more stars in the universe than grains of sands on earth.

Perhaps it is best to end our story with that touch of mystery. As cosmologist Carl Sagan asserted, "Science is not only compatible with spirituality; it is a profound source of spirituality." He was right. Most parents will tell you that there is no experience in life as transcendent as the birth of a child, with its mix of jubilation, humility, and awe. This was certainly the case for me. And yet my sense of wonderment—my spirituality—has only grown as my understanding of embryonic development has ripened.

As human beings, we tend to focus on our differences, which too easily blind us to our far-more-abundant similarities. That we all began unpretentiously, as a single cell, should be a source of solidarity, a reminder of our deep and irrevocable connections. Viewed from this vantage point, and embracing our shared origin, it is much easier to celebrate, rather than disparage, our differences.

"Every one of us is, in the cosmic perspective, precious," Sagan reminds us. "If a human disagrees with you, let him live. In a hundred billion galaxies, you will not find another."

ACKNOWLEDGMENTS

THIS BOOK IS ABOUT BEGINNINGS, SO IT IS FITTING THAT I START BY thanking the people without whom this project would have never gotten off the ground. Playwright and science enthusiast Deb Laufer perceived that the world needed to know more about embryonic development and convinced me, despite my misgivings, that I was the person to write about it. Al Zuckerman, my agent at Writers House, saw the idea's potential, and his encouragement gave me the confidence to bring the effort to fruition. I could not have asked for a better editor than Jessica Yao at Norton, whose comments and support led to a much improved and more readable manuscript. I am also grateful to Peter Klein, Jay Rajagopal, Yuval Dor, Vikram Paralkar, and Shannon Welch for their insightful comments on the manuscript at its various stages, and to Zovinar (Zovi) Khrimian, who did a wonderful job of demystifying difficult concepts and techniques with her illustrations.

David Housman, Phil Leder, and Doug Melton—my undergraduate, graduate, and postdoctoral advisers, respectively—taught me how to identify important questions, and I remain grateful for their passion and wisdom, which continue to inspire me. I started at the University of Pennsylvania in 2006, and from the day I set foot on campus, my colleagues have been endlessly supportive. Anil Rustgi and Celeste Simon drove the welcome wagon, helping me shape my laboratory into a productive and collaborative space. Bob Vonderheide, Ken Zaret, Klaus Kaestner, Ali Naji, Jon Epstein, Larry Jameson, Mike Parmacek, and others provided support in myriad ways, making the difficult job of running a research laboratory that much easier. Likewise, I am grateful to the members of my laboratory group, both past and present, whose curiosity and dedication to research perpetually renew my own enthusiasm for science. I am also indebted to Karen Miner for allowing me to share her story and to the many patients I have met during my career who, through their experiences (and their suffering), have taught me the preciousness of life, the power of science, and the importance of a sense of humor.

I owe much of the material in this book to the dozens of researchers whose discoveries have revealed so many of the embryo's secrets, and its potential. My attempt to construct a streamlined narrative meant, unfortunately, that I could recognize only a fraction of their contributions; I hope that they, and others in the field, will forgive these many omissions. For readers seeking a deeper and more comprehensive overview of the topic, I highly recommend Scott Gilbert's *Developmental Biology* textbook, which has been the bible for generations of budding developmental biologists.

Finally, I could not have done any of this without the loving support of my wife, Elsa, and our children, Sarah and Jacob. Elsa's patience and encouragement made this book possible.

GLOSSARY OF TERMS

adhesion complexes—groupings of ***adhesion molecules***, which associate (forming complexes) on the surface membranes of cells, especially ***epithelial*** cells. Adhesion complexes on one cell can bind to those present on adjacent cells, resulting in the tight cellular associations that make up an epithelium.

adhesion molecules—proteins located at or near the cell surface that associate to form ***adhesion complexes***.

alleles—variants in a *gene* (***genotype***) caused by differences in nucleotide sequence. A single gene may have one or more alleles.

angiogenesis—the formation of new blood vessels via sprouting from preexisting blood vessels, a feature of both growing embryos and growing tumors.

assay—an analytical procedure allowing the measurement of a molecular or biological property or outcome.

bacteriophage—See ***phage***.

basement membrane—a sheet-like layer of extracellular proteins and other macromolecules that lies beneath a layer of epithelial cells.

blastema—an aggregate of undifferentiated cells that forms at wound sites of certain organisms from which a completely new appendage can form, typically via ***epimorphic regeneration***.

blastoderm—the single layer of cells destined to form the embryo; referred to as the ***epiblast*** in vertebrates.

blastocyst—the mammalian equivalent of the ***blastula***, consisting of ***inner cell mass***, ***trophectoderm***, and blastocoel cavity.

blastomeres—the cells of the early embryo resulting from the ***cleavage*** divisions of the ***zygote***.

blastula—the stage of embryogenesis after the ***morula*** stage when the embryo consists of 100–150 cells; referred to as a ***blastocyst*** in mammals.

bone marrow transplantation—a technique in which bone marrow is transferred from a donor (animal or human being) to a recipient. The presence of ***hematopoietic stem cells*** in the bone marrow permits reconstitution of the blood system.

branching morphogenesis—the tissue-shaping process by which new tubular structures branch from a central trunk, like the branches of a tree. Examples include the lung, kidney, and mammary gland.

cancer stem cells (CSCs)—a hypothetical population of cells present within a tumor that fuels tumor growth (in contrast to the bulk of the tumor cells, which may have a more limited growth capacity).

cell fate—the default trajectory of a cell and its descendants during development and differentiation.

cell theory—the mid-nineteenth-century recognition that all living creatures are composed of cells.

cell therapy—the concept of delivering experimentally derived cells to an individual patient to treat disease.

cellular transplantation—the technique of moving cells from one region of an embryo to a similar or distinct region of the same or another embryo.

chimeric mouse—a mouse generated by combining blastomeres from two embryos or by introducing *stem cells* into a ***blastocyst***.

chromatin—the native state of genetic material inside the nucleus of ***eukaryotic*** organisms, consisting of a complex of DNA and protein (***histones***).

chromosomes—discrete packages of genetic material (***chromatin***) transmitted from a cell to its daughters with each round of cell division. ***Diploid*** cells have two sets of chromosomes, while ***haploid*** cells have a single set.

cleavage—the cell divisions of the early embryo whereby each daughter (***blastomere***) is half the size of its parent.

clonal—arising from a single cell or nucleus.

clone (cloning)—in organismal terms, the act of creating of a new organism from a single cell (for example, by ***nuclear transplantation***). In molecular biology terms, the isolation of a specific nucleic acid sequence that can then be manipulated.

codons—three-nucleotide codewords in DNA or RNA specific for particular amino acids.

colonies—discrete clusters of cells, typically ***clonal*** in origin.

commitment—the state of a cell whose identity or fate is fixed (in contrast to ***plasticity***).

compensatory growth—a means of tissue regeneration whereby the cells that remain after an injury divide to reconstitute normal tissue mass (in contrast to ***epimorphic regeneration***).

competence—the ability of a cell or its progeny to respond to certain signals or behave in a particular way. For example, ***commitment*** is associated with a loss of competence to adopt certain ***cell fates***.

conjugation—a bacterial mating process wherein "male" bacterial cells are able to share genetic material with "female" cells.

convergent extension—a morphogenetic process whereby cell tension causes tissues to elongate.

cytoplasm (cytoplasmic)—the part of the cell body distinct from the nucleus. It is the site of ***translation*** of mRNA into protein.

cytoskeleton—a network of intracellular proteins that give a cell its shape and allow it to move or deform.

de-differentiation—the process whereby a cell and/or its progeny loses its specialized properties.

diauxie—the sequential consumption of different sugars by bacterial cells.

differentiation—the process whereby a cell and/or its progeny acquires specialized properties.

diploid—a cell or organism having two copies of the ***genome***.

DNA methylation—a chemical modification to cytosine, one of the four DNA bases, resulting in the addition of a "methyl" group. Methylation of many cytosines surrounding a gene results in lower levels of ***transcription*** and gene ***expression***.

dominant—having the ability, as an allele, to override any other allele to confer a ***phenotype***.

E-cadherin—a protein component of membrane ***adhesion complexes*** that helps to promote the close association of adjacent epithelial cells.

ectoderm—one of the three embryonic ***germ layers*** whose descendants include the skin and the nervous system.

embryonal carcinoma cells (ECs)—cells derived from a teratoma having the property of ***multipotency***.

embryonic lethality—a ***phenotype*** resulting from the disruption of a gene that is required for embryonic development.

embryonic stem cells (ESCs)—cells derived from the ***inner cell mass*** of an embryo having the property of ***pluripotency***.

endoderm—one of the three embryonic ***germ layers*** whose descendants include the lungs, pancreas, liver, and intestines.

entelechy—term coined by Aristotle meant to refer to the "life force" or "soul" that impels the formation of living organisms.

epiblast—the single-layered state of the vertebrate embryo prior to ***gastrulation***. In invertebrate species, it is more often referred to as the ***blastoderm***.

epigenesis—the step-by-step development of an organism from an undifferentiated ***zygote*** (not to be confused with ***epigenetics***).

epigenetics—a mechanism of cellular memory that occurs without altering DNA sequences. In molecular terms, epigenetic regulation of differentiation is achieved by chemically modifying DNA (for example, ***DNA methylation***) or DNA-associated proteins (***histones***).

epigenome—the complete set of epigenetic modifications present in a cell or organism.

epimorphic regeneration—the regeneration of a tissue in its entirety from undifferentiated progenitors within a wound ***blastema***.

epithelial—a thin tissue layer lining a bodily surface (internal or external), forming a barrier.

eukaryotes—organisms or cells having a nucleus, distinguishing them from ***prokaryotes***.

expression (gene expression)—the degree to which a gene is ON by virtue of the level of ***transcription***.

galactosidase—an enzyme encoded by the *E. coli lacZ* gene permitting a cell to metabolize lactose as a fuel source.

gametes—the ***haploid*** cells of males and females (sperm and egg), which unite to form a ***zygote***.

gastrulation—a morphogenetic process marking the first major ***differentiation*** events in the embryo whereby the single-layered ***epiblast*** forms the three ***germ layers***: ***ectoderm***, ***endoderm***, and ***mesoderm***.

gemmules—hypothetical units of heredity proposed by Darwin to explain the inheritance of variant traits during evolution.

gene—a segment of DNA comprising a hereditary unit. A gene may encode a protein or some other functional product.

gene regulation—the molecular processes leading a gene to be turned ON (undergo ***transcription***) or OFF.

gene transfer—a method for getting cells to take up and incorporate foreign pieces of DNA.

genetic carrier—an animal capable of passing along a mutant allele to its offspring but which itself lacks any phenotype (typically occurs when a ***recessive*** mutant allele is present in a ***heterozygous*** state).

genetic screen—a method in which mutants are evaluated to identify a subset with important roles in development or normal physiology.

genetics (genetic approach)—a method in which mutants are evaluated to identify a subset causing a particular phenotype, thereby revealing roles in development or normal physiology.

genome—the complete set of genes in an organism.

genomic equivalence—the concept that all cells within an organism carry the same ***genome***.

genotype—a particular combination of ***alleles*** for a given ***gene*** or ***genes***.

germ layers—the three layers of cells (***ectoderm***, ***endoderm***, and ***mesoderm***) arising after ***gastrulation*** from which all organs and tissues arise in an animal.

germ plasm theory—the nineteenth-century hypothesis advanced by August Weismann which stated that cells lose determinants (***genes***) as they ***differentiate*** such that a cell's specialized identity results from whatever genes remain.

haploid—a cell or organism having one copy of the ***genome***.

hematopoiesis—the process of blood cell production and differentiation.

hematopoietic stem cell (HSC)—a cell which normally resides in the bone marrow that can reconstitute all the cellular lineages of the blood.

heterozygous—having two different alleles comprising a ***diploid genotype***.

histones—proteins found in association with DNA to form ***chromatin***.

homeotic mutation—a mutation leading to the substitution of one body part for another.

homologous recombination—the process by which a cell swaps one segment of DNA for another based on a close similarity of the two DNA sequences.

homozygous—having two identical alleles comprising a ***diploid genotype***.

hypertrophy—growth of a tissue or cell (as opposed to an increase in cell number).

induced pluripotent stem cells (iPSCs)—stem cells derived by converting (***reprogramming***) specialized cells to a ***pluripotent*** state.

induction—(1) in the context of ***phage***, causing the virus to awake from its ***lysogenic cycle*** and enter a ***lytic cycle***; (2) in the context of ***gene regulation***, the process of turning a gene ON by activating its ***transcription***; (3) in development, the process whereby certain cells influence the behavior of other cells.

inner cell mass (ICM)—the cluster of cells in a ***blastocyst*** that will give rise to the embryo.

intravasation—the process by which a cancer cell enters the bloodstream.

invasion—the process by which a cell moves into deeper tissue layers.

invertebrate—any animal lacking a backbone. Examples include snails, insects, corals, worms, octopuses, jellyfish, and so on.

lysogenic cycle—dormant portion of the ***phage*** life cycle when the virus "sleeps" in its bacterial host.

lytic cycle—active portion of the ***phage*** life cycle when the virus replicates and ultimately ruptures ("lyses") its bacterial host.

mesenchymal—resembling or derived from the ***mesoderm*** germ layer.

mesoderm—one of the three embryonic ***germ layers*** whose descendants include bone, muscle, and cartilage.

metastasis—the phenomenon of tumor cell spread from its primary organ to distant sites.

micrometastasis—early stage of ***metastasis*** when tumor cell spread is undetectable by standard clinical means.

morphogenesis—the acquisition of tissue shape.

morula—an early stage of development when the embryo is a ball of roughly 16–32 cells.

mosaic model—the centerpiece of the ***germ plasm theory*** which hypothesized that fate-determining factors are unequally distributed within the egg such that each segment of the egg gives rise to a different part of the future animal.

multipotency—the ability of a cell to give rise to multiple types of specialized progeny.

natural selection—the centerpiece of Darwin's theory of evolution whereby any heritable variation that increases the fitness of an animal becomes overrepresented in the population.

nephron—the functional filtration unit of the kidney.

niches—specialized regions that help tissue ***stem cells*** retain their unique properties.

nuclear transplantation—the method of isolating a ***nucleus*** from one cell and inserting it into the enucleated ***cytoplasm*** of another cell.

nucleosomes—repeating units of ***chromatin*** consisting of double-stranded DNA wrapped around cores of ***histone*** proteins.

nucleus (nuclear)—the part of the cell containing the genetic material; the site of ***transcription*** of DNA into mRNA.

organizer—a group of cells having the ability to instruct fate and morphogenesis in adjacent cells through ***induction***.

pangenesis—Darwin's theory of inheritance based on the idea that circulating ***gemmules*** are passed on to an animal's progeny.

patterning—the process of giving spatial organization to the embryo during development.

permease—a protein channel (encoded by the *lacY* gene in *E. coli*) that permits lactose to enter the bacterial cell.

phage (bacteriophage)—a virus or viruses capable of infecting bacterial cells.

phagocytosis—the process by which one cell engulfs another cell.

phenotype—the cellular or animal trait manifested by ***genotype***.

planar cell polarity—a phenomenon whereby cells within an epithelial layer can distinguish front from back.

plasticity—the ability of cells to change their developmental trajectories or differentiated identities.

pluripotent—able to give rise to all the cell types of an animal.

polarity—the cellular asymmetries that arise from a cell existing in three-dimensional space. For example, apical-basal polarity describes the difference between a cell's top and its bottom, while ***planar cell polarity*** describes the difference between front and back.

polyclonal—arising from many cells.

positional identity—the property by which cells recognize their position in three-dimensional space relative to other cells.

preformationism—the anachronistic notion that the body exists in a preformed state inside the egg or sperm.

progenitor cell—an undifferentiated cell with the potential to give rise to other differentiated cells (but whose potential is less than that of a ***stem cell***).

prokaryotic—lacking a nucleus.

prophage—the dormant or ***lysogenic*** state of a ***phage***.

recessive—having the ability, as an allele, to confer a phenotype only when unopposed by a coexisting ***dominant*** allele.

repressor—See ***transcriptional repressor***.

reprogramming—the molecular remodeling of a cell, causing it to take on a new identity.

reverse genetics—a method in which a known ***gene*** (or ***genes***) is evaluated for its ability to confer a ***phenotype***, thereby revealing its roles in development or normal physiology.

stem cell—any single cell with the properties of ***multipotency*** (the ability to differentiate into other types of cells) and self-renewal (the ability to make more stem cells). In adult stem cells, potential is limited to the cell types normally present in the tissue where they reside. In ***embryonic stem cells (ESCs)***, potential extends to the entire animal (***pluripotency***).

synapses—the connections between neurons that allow the nervous system to form an integrated network.

syncytium—a cell containing multiple nuclei.

teratoma—a tumor arising from the gonads (ovaries or testes) containing cells with features of multiple differentiated lineages.

transcription—the production of messenger RNA (mRNA) copies from a DNA template, used to control the level of ***gene expression***.

transcription factor—a DNA-binding protein that regulates the level of ***gene expression*** either positively (***transcriptional activator***) or negatively (***transcriptional repressor***).

transcriptional activator (activator)—a DNA-binding protein that facilitates ***transcription***.

transcriptional repressor (repressor)—a DNA-binding protein that inhibits ***transcription*** (the lactose repressor is encoded by the *lacI* gene in *E. coli*).

transgenerational inheritance—transmission of an acquired trait from an individual to their offspring.

transgenic mouse—a genetically engineered mouse into which an exogenous piece of DNA has been incorporated into the ***genome***.

translation—the conversion of an mRNA template into its protein product.

trophectoderm—the cells lining the outer surface of a ***blastocyst*** that will give rise to the placenta.

tumor microenvironment—the cellular community within a tumor consisting of cancer cells as well as noncancer cells (fibroblasts, immune cells, and blood vessels).

white—a ***gene*** in fruit flies which, mutated, causes the flies to have white eyes instead of their normally red eyes. It was the first animal mutation to be identified in a laboratory.

xenotransplantation—the transplantation of cells from one species into another species. Can apply to organs (for example, the transplantation of pig organs into human beings) or cancers (for example, the transplantation of human cancer cells into mice).

zygote—the single-celled embryo formed by fusion of egg and sperm.

NOTES, REFERENCES, AND FURTHER READING

PROLOGUE: CONCEPTION

5 **each equation he included:** StephenHawking, *A Brief History of Time* (New York: Bantam Books, 1988), vi.

CHAPTER 1: THE ONE CELL PROBLEM

10 **Aristotle's logic silenced the preformationists:** Aristotle conveniently sidestepped the question of how those parts came to be assembled, invoking invisible forces that he suggested were generated when sperm and egg joined—forces he referred to as "life" or "soul."

12 **a reexamination of Aristotle's methods:** Aristotle had fallen out of favor with many thinkers by this time—his methods of inquiry were seen as stale—which further set the stage for countervailing views to be readily embraced. See Clara Pinto-Correia's *The Ovary of Eve* (Chicago: University of Chicago Press, 1997), 25.

12 **waiting for some signal to unfold:** In retrospect, Swammerdam's observations were correct, but his interpretation was wrong. Swammerdam was examining insects during "metamorphosis," a developmental stage just before the adult insect is born. At this stage, many parts of the body have indeed already formed and simply expand when the adult emerges from its pupal shell or cocoon.

13 **the testes of Adam:** Pinto-Correia, *The Ovary of Eve*, 65.

14 **Darwin's radical idea:** Charles Darwin, *On the Origin of Species by Means of Natural Selection* (London: John Murray, 1859).

16 **The cell divisions that occur:** Walther Flemming, another German scientist, was the first to perform a detailed analysis of cell division. Flemming discovered

that when cells were treated with certain dyes, different subcellular structures—organelles—became easier to see. His observations of chromosome behavior during cell division resulted in a detailed description of the process, including the phases of mitosis (prophase, metaphase, anaphase, and telophase) that we are taught in high school biology.

18 **determinants that he called the *germ plasm*:** August Weismann, *The Germ-Plasm: A Theory of Heredity*, trans. W. Newton Parker and Harriet Ronnfeldt (New York: Scribner, 1893).

18 **"If the egg were truly unorganized":** Stephen Jay Gould, *Ever Since Darwin* (New York: W. W. Norton, 1992), 205.

21 **oblivious to their neighbors' fate:** Wilhelm Roux, "Contributions to the Developmental Mechanics of the Embryo. On the Artificial Production of Half-Embryos by Destruction of One of the First Two Blastomeres, and the Later Development (Postgeneration) of the Missing Half of the Body" (1888), in *Foundations of Experimental Embryology*, ed. Benjamin Willier and Jane Oppenheimer (Englewood Cliffs, NJ: Prentice-Hall, 1964).

21 **each made a unique contribution:** Laurent Chabry, "Contribution à l'embryologie normale tératologique des ascidies simples," *Journal de l'anatomie et de la physiologie normales et pathologiques de l'homme et des animaux* 23 (1887): 167–321.

24 **an almost unbelievable result:** Hans Driesch, "The Pluripotency of the First Two Cleavage Cells in Echinoderm Development. Experimental Production of Partial and Double Formations" (1892), in *Foundations of Experimental Embryology*, ed. Benjamin Willier and Jane Oppenheimer (Englewood Cliffs, NJ: Prentice-Hall, 1964).

26 **what if this assumption was false:** When American researcher Jessie McClendon later used Driesch's approach to test the potential of frog blastomeres, separating them instead of killing them, the frog cells behaved just like sea urchin cells, proving that the difference in species was not responsible for the different results. Jessie Francis McClendon, "The Development of Isolated Blastomeres of the Frog's Egg," *American Journal of Anatomy* 10 (1910): 425–30.

26 **hemi-embryos instead of complete ones:** Driesch's results were also inconsistent with those of Chabry, the French scientist who had separated blastomeres from invertebrate sea squids. Unlike Driesch's blastomeres, which went on to form a new animal, those of Chabry behaved autonomously. This discrepancy, unlike the one involving Roux, is related to the different species each scientist used—a difference that proved critical. Some animals—particularly invertebrates like the sea squirt and certain mollusks—develop in an autonomous fashion at early stages of embryogenesis. Such behavior is rare in vertebrates and, as it turned out, does not drive sea urchin development either. Nevertheless, an interplay between determinism and plasticity is present to some degree in all animal development, with the balance shifting as a function of time, species, and environment.

28 **a knack for working with fragile specimens:** This description of Spemann's transplantation experiments skims the surface of his work and contributions to embryology. His earliest experiments involved manipulations of much younger embryos, but they were performed with the same adeptness. One of his most famous experiments involved the constriction (partial division) of early newt embryos using one of his daughter's hairs as a ligature.

32 **the *organizer*:** Spemann, along with American embryologist Walter Lewis, had seen evidence for induction decades earlier while studying eye development in frogs. Although the Spemann-Mangold organizer experiment is the most famous and dramatic example of induction, there are many other tissues throughout development that serve as organizing centers—groups of cells with the capacity to alter fate and create organized structures. See Viktor Hamburger, *The Heritage of Experimental Embryology: Hans Spemann and the Organizer* (New York: Oxford University Press, 1988).

32 **Her paper had just come out:** Hans Spemann and Hilde Mangold, "Induction of Embryonic Primordia by Implantation of Organizers from a Different Species (1924)," trans. Viktor Hamburger, *International Journal of Developmental Biology* 45 (2001): 13–38.

CHAPTER 2: THE LANGUAGE OF THE CELL

36 **"patiently accumulating":** Darwin, *Origin of Species*, 1.

38 **he, too, believed in blending:** Brian Charlesworth and Deborah Charlesworth, "Darwin and Genetics," *Genetics* 183 (November 2009): 757–66.

39 **Weismann examined more than 900 mice:** August Weismann, "The Supposed Transmission of Mutilations," chap. 8 in *Essays upon Heredity and Kindred Biological Problems*, trans. and ed. Edward B. Poulton, Selmar Schonland, and Arthur E. Shipley (Oxford: Clarendon Press, 1889).

39 **Darwin came to recognize the flaws in his own model:** Charles Darwin, *The Variation of Animals and Plants under Domestication* (London: John Murray, 1868).

40 ***per scientiam ad sapientiam*:** Robin Marantz Henig, *The Monk in the Garden: The Lost and Found Genius of Gregor Mendel* (Boston: Mariner Books, 2000), 23.

41 **an end to the experiments:** Kenneth Paigen, "One Hundred Years of Mouse Genetics: An Intellectual History," *Genetics* 163 (April 2003): 1.

41 **plants also have sex:** Henig, *Monk in the Garden*, 16.

44 **only a single, disparaging, reply:** Much has been written about the extent to which Mendel and Darwin knew about each other's work. Mendel owned a copy of *On the Origin of Species* (second German edition, published 1863), and had marked numerous passages of interest. Mendel's paper never mentions Darwin by name, perhaps to avoid the appearance of confrontation with the famous biologist. Going further, Mendel may have even intended for his form-building elements to serve as a solution to the glitch in Darwin's theory—What was the source of variation on which natural selection could act?—a question that the notion of blending could not solve. Regarding Darwin's knowledge of Mendel, the answer is less clear. Mendel ordered 40 or so reprints of his paper, which he sent to notables across Europe. Recipients included several scientists we will meet later in the book, including Theodor Boveri and Matthias Schleiden. Darwin would have certainly been on his list. Even so, Darwin may have given the manuscript only cursory attention, given its emphasis on numbers (Darwin allegedly said that "mathematics in biology was like a scalpel in a carpenter's shop—there was no use for it"). For further reading, see David Galton, *Standing on the Shoulders of Darwin and Mendel: Early Views of Inheritance* (Boca Raton, FL: Taylor and Francis Group, 2018); Gavin de Beer, "Mendel, Darwin, and Fisher (1865–1965)," *Notes and Records of the Royal Society of London* 19, no. 2

(December 1964): 192–226; Hub Zwart, "Pea Stories: Why Was Mendel's Research Ignored in 1866 and Rediscovered in 1900?," in *Understanding Nature: Case Studies in Comparative Epistemology* (Dordrecht, Netherlands: Springer, 2014), 197.

46 **what caught his attention:** Elof Carlson, "How Fruit Flies Came to Launch the Chromosome Theory of Heredity," *Mutation Research* 753, no. 1 (July–September 2013):1–6.

46 **Morgan had other ideas:** Despite Morgan's lack of interest, Payne went ahead and bred flies in the dark for 49 generations, corresponding to 15 centuries on a human scale. As we might have predicted, he found no evidence that the flies lost the ability to see.

47 **"And how is the baby?":** Garland Allen, *Thomas Hunt Morgan: The Man and His Science* (Princeton, NJ: Princeton University Press, 1978), 153.

47 **the nickname "*white*":** Geneticists typically refer to a gene and to animals bearing mutant versions of the gene with a term that reflects the mutant phenotype. In this case, the gene responsible (when mutated) for giving rise to white-eyed flies was called *white*; mutant animals bearing mutations in the *white* gene are thus referred to by the shorthand "*white* flies."

50 **those "chromosomal people":** Allen, *Thomas Hunt Morgan*, 139.

50 **Stevens correctly theorized:** Sarah Carey, Laramie Akozbek, and Alex Harkness, "The Contributions of Nettie Stevens to the Field of Sex Chromosome Biology," *Philosophical Transactions of the Royal Society B* 377 (2022): 1–10.

51 **Morgan continued his breeding:** Following the logic above, it is possible to see how Morgan obtained white-eyed females through additional breeding. Specifically, by crossing a female fly carrying one mutant *white* allele (X^wX^+) with a white-eyed male (X^wY), half of the females would have white eyes (X^wX^w) and half would have red eyes (X^wX^+).

51 **Genetics, as the nascent field:** The terms "gene" and "genetics"—coined by Wilhelm Johannsen and William Bateson, respectively, as a simple way to refer to Mendel's form-building elements—are derived from the Greek word *genos* ("race" or "stock").

52 **a physical reconfiguration of the chromosome:** The model of genes "traveling together" received further support when Belgian biologist Frans Janssens discovered that cells can swap bits of chromosomes during the formation of gametes—a process he called "crossing over." This was direct evidence that car-swapping occurs between pairs of chromosomes.

53 **these traits were too close together:** To give a bit more granularity to this specific example, consider the vermillion eye trait to be encoded by the *vermillion* gene, with its two alleles, V (wild-type dominant) and v (mutant recessive), and consider the miniature wing trait to be encoded by the *miniature* gene, with its two alleles, M (wild-type dominant) and m (mutant recessive). By this definition, the original (purebred) stocks of vermillion flies would have the genotype vv/MM (mutant for the eye trait but wild-type for the wing trait) while the original stocks of the miniature winged flies would have the genotype VV/mm (wild-type for the eye trait but mutant for wing trait). As the genes encoding these two traits are tightly linked, this means that in the vermillion flies, a v allele is located near an M allele; in the miniature flies, a V allele is located near an m allele. To generate flies with vermillion eyes and miniature wings—with the genotype vv/mm—the mutant alleles would have to become untethered from the wild-type alleles; the only way for this to happen would be through a crossing-over (car-swapping) event.

53 **the first "genetic map":** The unit of a genetic map, still in use today, is the "centimorgan," which reflects a 1 percent chance of two genes becoming separated from one generation to the next. In humans, a distance between two genes of one centimorgan corresponds to approximately one million bases of DNA. Alfred Sturtevant, "The Linear Arrangement of Six Sex-Linked Factors in Drosophila, as Shown by Their Mode of Association," *Journal of Experimental Zoology* 14 (1913): 43–59.

55 **the nucleus remained intact:** Ralf Dahm, "Friedrich Miescher and the Discovery of DNA," *Developmental Biology* 278, no. 2 (2005): 274–88.

55 **vastly different forms:** Water molecules (H_2O) have a "U" shape, while hydrogen peroxide molecules (H_2O_2) are shaped like the twisted handlebars from a damaged bicycle. As compounds become larger, their structures typically become harder to solve.

56 **the obvious candidate for the genetic material:** Although the identification of DNA as the hereditary material would not formally occur before the 1940s, developmental biologist (and friend of Hans Driesch) Oskar Hertwig wrote in 1885, "Nuclein is the substance responsible not only for fertilization but also for the transmission of hereditary characteristics." But few paid attention. John Gribben, *The Scientists: A History of Science Told through the Lives of Its Greatest Inventors* (New York: Random House, 2004), 547.

57 **three million possible combinations:** The combinatorial encoding power of proteins derives from the order of amino acids at each position. In a protein containing 5 amino acids, for example, any of 20 amino acids could occupy the first position, 20 the second position, etc. Hence, the number of permutations in a protein containing only 5 amino acids is $20 \times 20 \times 20 \times 20 \times 20 = 3{,}200{,}000$ (and most proteins contain several hundred amino acids). By comparison, the combinatorial power of five DNA "bases" strung together is much lower, since there are only four bases to choose from (A, G, C, or T) in each of the five positions. Thus, in a comparable nucleic acid containing five bases, there are only a thousand possible combinations (i.e., $4 \times 4 \times 4 \times 4 \times 4 = 1{,}024$).

57 **which typically killed the host:** It turns out that the difference in lethality is due to differences in the ability of the immune system to recognize and clear the two bacterial strains. Bacteria of the R-form have a rough outer shell, making them easier for immune cells to detect and clear, resulting in a benign course. Bacteria of the S-form, by contrast, have a poorly recognized smooth outer shell, providing them with a greater ability to evade the immune system, and hence more severe disease.

58 **"principally, if not solely":** Oswald Avery, Colin MacLeod, and Maclyn McCarty, "Studies on the Chemical Nature of the Substance Inducing Transformation of Pneumococcal Types," *Journal of Experimental Medicine* 79, no. 2 (1944): 137–58.

CHAPTER 3: SOCIETIES OF CELLS

68 **Gurdon found himself dead last:** "Sir John B. Gurdon: Biographical," Nobel Prize.org, Nobel Prize Outreach AB 2022, accessed October 22, 2022, https://www.nobelprize.org/prizes/medicine/2012/gurdon/biographical/.

70 **picture an omniscient librarian:** You may wonder, in this metaphor of the library, who the librarian is and how they are deciding which books to remove and which should remain. While Weismann's theory proved to be incorrect (no genes are removed during development), the analogy still holds with respect to

gene regulation—which genes remain ON or are turned OFF during development (the focus of Chapter 4)—where the answer is still mysterious. The instructions leading to cellular specialization are embedded in our genomes, but we have only a rudimentary idea of how these instructions are so reproducibly parsed to different cells during development.

72 **a "fantastical experiment":** Hans Spemann, *Embryonic Development and Induction* (New Haven, CT: Yale University Press, 1938).

72 **believed he could do it:** It is unclear whether Briggs knew of Spemann's suggestion, but it appears likely that he initially pursued this experiment unaware of any historical precedence.

72 **a third of the hybrids grew:** Robert Briggs and Thomas King, "Transplantation of Living Nuclei from Blastula Cells into Enucleated Frogs' Eggs," *Proceedings of the National Academy of Sciences USA* 38 (May 1952): 455–63.

76 **half as big as its parent:** For a beautiful picture of cleavage in the African clawed frog, *Xenopus laevis*, see H. Williams and J. Smith, Xenopus laevis *Single Cell to Gastrula*, video posted by xenbasemod October 21, 2010, YouTube, http://www.youtube.com/watch?v=IjyemX7C_8U.

79 **the odds of success dropped to zero:** Thomas King and Robert Briggs, "Serial Transplantation of Embryonic Nuclei," *Cold Spring Harbor Symposia on Quantitative Biology* 21 (1956): 271–90; Robert Briggs and Thomas King, "Changes in the Nuclei of Differentiating Endoderm Cells as Revealed by Nuclear Transplantation," *Journal of Morphology* 100 (March 1957): 269–312.

81 **everything would work out:** Perhaps this overstates Gurdon's risk. When he began the experiments, and didn't know what result he would get, there was a decent chance that he would simply find the same thing Briggs and King had found, namely, that differentiated nuclei were incapable of supporting development. If that turned out to be the finding, then it would raise the question of how genes were lost during development and how the embryo decided which cells would lose which genes. So in the end, either result could have led to another productive line of investigation (at least in principle).

83 **"Could there be an artifact":** John Gurdon, "Revolution in the Biological Sciences," interview by Harry Kreisler, Conversations with History, Institute of International Studies, University of California, Berkeley, https://conversations.berkeley.edu/gurdon_2006.

84 **easily distinguishable under the microscope:** The mutation in this strain of *Xenopus* caused it to have only one nucleolus instead of the normal two, but it had no effect on the health, fertility, or development of the frogs. The mutant phenotype was noted by one of Fischberg's other students, Sheila Smith, who happened upon the outlier. Fischberg asked her to locate and recover the mutation-bearing frog, whose descendants were then used for Gurdon's subsequent cloning experiments. T. R. Elsdale, M. Fischberg, and S. Smith, "A Mutation That Reduces Nucleolar Number in *Xenopus laevis*," *Experimental Cell Research* 14, no. 3 (1958): 642–43.

85 **spawn a new animal:** M. Fischberg, J. B. Gurdon, and T. R. Elsdale, "Nuclear Transplantation in *Xenopus laevis*," *Nature* 181 (February 1958): 424; M. Fischberg, J. B. Gurdon, and T. R. Elsdale, "Sexually Mature Individuals of *Xenopus laevis* from the Transplantation of Single Somatic Nuclei," *Nature* 182 (July 1958): 64–65.

86 **more mature nuclear donors:** John Gurdon, "The Developmental Capacity of Nuclei Taken from Intestinal Epithelium Cells of Feeding Tadpoles," *Journal of*

Embryology and Experimental Morphology 10 (December 1962): 622–40; J. Gurdon and V. Uehlinger, "'Fertile' Intestine Nuclei," *Nature* 210 (June 1966): 1240–41; Ronald Laskey and John Gurdon, "Genetic Content of Adult Somatic Cells Tested by Nuclear Transplantation from Cultured Cells," *Nature* 228 (December 1970): 1332–34.

86 ***Brave New World*:** Aldous Huxley, *Brave New World* (London: Chatto and Windus Press, 1932).

86 **to a different level:** The reasons for the discrepancy between Gurdon's results and those of Briggs and King (i.e., why he succeeded where they failed) are not entirely clear. It may be that technical differences render *Rana* nuclei less susceptible to the rejuvenating effects of the egg. Ultimately, successful nuclear transplantation was achieved in *Rana pipiens*, demonstrating that genomic equivalence operates in this strain as well. See Nancy Hoffner and Marie DiBerardino, "Developmental Potential of Somatic Nuclei Transplanted into Meiotic Oocytes of *Rana pipiens*," *Science* 209 (July 1980): 517–19.

87 **the genetic instructions needed:** In our discussion so far, I have neglected one important, if not obvious, point: nuclear transplantation worked only if an egg, and not any other type of cell, served as the recipient of the relocated nucleus. Something in the *cytoplasm* of the egg—the material comprising the egg's body minus its nucleus—was responsible for setting back the developmental clock of its new guest. This fact points to an interplay between nucleus and cytoplasm that is particularly critical during early development. (The molecular identity of this rejuvenating factor is still unknown.) For a further discussion of this complex topic, see Michael Barresi and Scott Gilbert, *Developmental Biology*, 12th ed. (New York: Sinauer Associates, 2020).

88 **Dolly was the result:** I. Wilmut, A. E. Schnieke, J. McWhir, A. J. Kind, and K. H. Campbell, "Viable Offspring Derived from Fetal and Adult Mammalian Cells," *Nature* 385 (February 1997): 810–13.

88 **cloned human babies:** Several groups have succeeded in using nuclear transfer to create cloned human embryos; in all these studies, embryos were not allowed to develop beyond the blastula stage. See Andrew French, Catharine Adams, Linda Anderson, John Kitchen, Marcus Hughes, and Samuel Wood, "Development of Human Cloned Blastocysts Following Somatic Cell Nuclear Transfer with Adult Fibroblasts," *Stem Cells* 26 (February 2008): 485–93; Scott Noggle, Ho-Lim Fung, Athurva Gore, et al., "Human Oocytes Reprogram Somatic Cells to a Pluripotent State," *Nature* 478 (October 2011): 70–75; Masahito Tachibana, Paula Amato, Michelle Sparman, et al., "Human Embryonic Stem Cells Derived by Somatic Cell Nuclear Transfer," *Cell* 153 (June 2013): 1228–38.

89 **"Someday, if I don't have a laboratory":** Conversations of the author with John Gurdon, April 2012.

CHAPTER 4: TURNING GENES ON AND OFF

93 **leaping headfirst into his future:** François Jacob, *The Statue Within*, trans. Franklin Philip (New York: Basic Books, 1988), 213.

97 **the moment that everyone would clear out:** Jacob, *The Statue Within*, 232–33.

99 **the order of those genes:** As an analogy, imagine a line of people holding hands and passing through a doorway one by one. If we close the door after 5 seconds, we

find that Jane is the only one on the other side. If we start over and then close the door after 10 seconds, both Jane and Patrick have made it through. And if we start again and wait 15 seconds before closing the door, we find Jane, Patrick, and Eve on the other side of the doorway. From this, we can infer that the line consisted of Jane followed by Patrick followed by Eve.

99 **bacterial coitus interruptus:** According to legend, Wollman purchased the gift and then promptly "borrowed" it for his experiments, as the laboratory had insufficient funds to purchase a blender, an appliance that was still a rarity in Europe.

102 **chopped the lactose:** Deficiencies in lactase, an enzyme made by the human intestine with similar function, is responsible for the human syndrome of lactose intolerance.

105 **a chemical analog:** The new reagent was a chemical called o-nitrophenyl-beta-galactopyranoside, or ONPG, which could be acted upon by the galactosidase enzyme. Chemicals that can be modified by an enzyme are known as enzyme substrates. When galactosidase acted on the ONPG substrate, it caused the substance to turn yellow, providing a quantitative readout for galactosidase enzyme activity.

106 **galactosidase production ceased:** Arthur Pardee, François Jacob, and Jacques Monod, "The Genetic Control and Cytoplasmic Expression of 'Inducibility' in the Synthesis of β-Galactosidase by *E. coli*," *Journal of Molecular Biology* 1 (June 1959): 165–78.

107 **one-to-one relationship with proteins:** George Beadle and Edward Tatum, "Genetic Control of Biochemical Reactions in Neurospora," *Proceedings of the National Academy of Sciences USA* 27, no. 11 (November 1941): 499–506.

108 **already a celebrated biologist:** Watson and Crick shared the Nobel Prize with Maurice Wilkins in 1962 "for their discoveries concerning the molecular structure of nucleic acids and its significance for information transfer in living material." The award famously failed to cite the fact that Rosalind Franklin, a structural biologist at King's College London, had performed the X-ray studies that proved crucial for the elucidation of the double-helical structure. Sadly, Franklin died of ovarian cancer in 1958, cutting short what would certainly have been an extraordinary career.

109 **the total RNA in a cell:** Cells contain several classes of RNA. The most abundant are so-called transfer RNAs (tRNAs) and ribosomal RNAs (rRNAs). These RNA molecules are involved in the production of a protein product from a DNA template, acting at the final step, translation. By contrast, messenger RNA (mRNA) is a tiny fraction of the total RNA in a cell. Yet it is the sole go-between linking DNA to protein, allowing the information embedded in a single gene to be amplified into numerous protein molecules.

112 **untouchable nature:** Jacob, *The Statue Within*, 302.

113 **a near mirror image:** One small difference between DNA and RNA is the sugar component of the base. Instead of being composed of *deoxyribonucleotides*, the building block of DNA, mRNA is built from related but functionally distinct *ribonucleotides*. This difference makes mRNA far less stable than DNA. Another difference is that mRNA molecules are single stranded; consequently, the proteins they encode are ultimately traceable back to only one of the two DNA strands making up a gene.

113 **RNA polymerase is free to act:** François Jacob and Jacques Monod, "Genetic Regulatory Mechanisms in the Synthesis of Proteins," *Journal of Molecular Biology* 3, no. 3 (June 1961): 318–56.

116 **a linguistic overhaul:** In addition to the regulation of transcription and translation, which determine the extent to which genes are expressed as their mRNA and protein products, respectively, proteins can be chemically modified even after they are synthesized. These modifications—glycosylation, phosphorylation, ubiquitination, acetylation, methylation—involve the addition of other chemical groups to a protein's amino acid backbone, affecting the protein's activity, stability, or position in the cell. Such "post-translational modifications," as they are called, are beyond the scope of this book. For further reading, see Harvey Lodish, Arnold Berk, Chris Kaiser, et al., *Molecular Cell Biology*, 9th ed. (New York: W. H. Freeman, 2021).

CHAPTER 5: GENES AND DEVELOPMENT

122 **"decidedly lazy":** "Christiane Nüsslein-Volhard: Nobel Prize in Physiology or Medicine 1995," NobelPrize.org, accessed October 22, 2022, https://www.nobelprize.org/womenwhochangedscience/stories/christiane-nusslein-volhard.

124 **known as the syncytial *blastoderm*:** Different terms are used to describe comparable structures in different species. For example, "blastocyst" and "blastula" refer to a similar point in development prior to gastrulation in mammalian and non-mammalian species, respectively. In insects, "blastoderm" refers to the single layer of embryonic cells that will subsequently undergo gastrulation; it is the equivalent of the mammalian "epiblast," which we will hear more about later.

130 **120 genes:** The Heidelberg scientists began with approximately 27,000 strains of mutagenized flies, out of which 18,000 were embryonic lethal. Scanning through the latter mutants one at a time, they identified approximately 600 that had a recognizable disturbance in form. These 600 strains, in turn, corresponded to 120 genes, meaning that the genes identified by the screen had each been mutated an average of four times. (One can tell whether multiple mutant strains correspond to the same gene by determining the chromosomal location of each mutation via chromosomal mapping, the technique Morgan and Sturtevant used decades earlier; if multiple mutants map to the same chromosomal location, they likely represent independent mutations in the same gene.) The fact that most genes were identified more than once provided further reassurance that the screen had captured as many patterning genes as it possibly could have (what is termed "saturation" of the screen). See Christiane Nüsslein-Volhard and Eric Wieschaus, "Mutations Affecting Segment Number and Polarity in *Drosophila*," Nature 287 (1980): 795–801.

130 **how long it takes to build a *Drosophila* larva:** It takes only one day for the *Drosophila* to reach the larval stage following fertilization of the egg. While the Heidelberg screen captured most of the genes involved in establishing the embryo's pattern, a limitation of the genetic approach is gene "redundancy," which occurs when more than one gene can perform a given function. When this occurs, each gene can act as a stand-in for the other, precluding mutant phenotypes from appearing. Such redundancy is common among vertebrates. But fortunately for *Drosophila* researchers, gene redundancy is comparatively rare in the fly.

132 **as simple as it got:** Sydney Brenner, "Nature's Gift to Science" (Nobel lecture, Stockholm, December 8, 2002), NobelPrize.org, accessed October 22, 2022, https://www.nobelprize.org/uploads/2018/06/brenner-lecture.pdf.

132 **one major difference:** A minor difference between Brenner's approach and that

of the Heidelberg scientists involved breeding. Unlike *Drosophila*, most *C. elegans* worms are hermaphrodites, containing both male and female sex organs. Although all of Mendel's laws of genetics still apply, genetic studies in *C. elegans* require some additional considerations.

133 **barely a ripple:** Sydney Brenner, "The Genetics of *Caenorhabditis elegans*," Genetics 77, no. 1 (May 1974): 71–94.

134 **"There's little point":** John Sulston, "*C. elegans*: The Cell Lineage and Beyond" (Nobel lecture, Stockholm, December 8, 2002), NobelPrize.org, accessed October 22, 2022, https://www.nobelprize.org/uploads/2018/06/sulston-lecture.pdf.

134 **1/40th the size:** John Sulston and Sydney Brenner, "The DNA of *Caenorhabditis elegans*," *Genetics* 77, no. 1 (May 1974): 95–104.

136 **early stages of invertebrate development:** Edwin Conklin, "Organization and Cell Lineage of the Ascidian Egg," *Journal of the Academy of Natural Sciences of Philadelphia*, 2nd ser., vol. 13, pt. 1 (1905).

137 **little tolerance for deviation:** The variation in cell number also depends on the sex of the worm. Hermaphrodites, which contain both male and female sex organs, have 959 cells, while males have 1,033 cells. Environmental influences in the adult may also have an impact, increasing or decreasing the number of cells by two or three.

137 **Nature's method for building an animal:** Previously, we described embryonic development as largely plastic—regulated by the signals cells receive from their neighbors. The developing worm shares these properties, but it also exhibits features of mosaic development, wherein cells follow an autonomous trajectory. If, for example, one separates the blastomeres of *C. elegans* at the two-cell stage, the equivalent of Driesch's shaking of the two-celled sea urchin embryo, one cell develops as it would have if it were left intact, giving rise to only the posterior half of the worm.

137 **until development was complete:** J. E. Sulston, E. Schierenberg, J. G. White, and J. N. Thomson, "The Embryonic Lineage of the Nematode *Caenorhabditis elegans*," *Developmental Biology* 100, no. 1 (November 1983): 64–119.

138 **recombinant DNA technology:** The molecular toolbox known as recombinant DNA technology emerged in the late 1970s. By exploiting *E. coli* and its bacteriophage, lambda (λ), scientists developed methods for isolating specific fragments of DNA (a process known as "molecular cloning"), making millions of copies, and determining the underlying sequences. While bacteria and phage have remained the vehicles for working with DNA fragments, nucleic acids from any species can now be subjected to the same kinds of molecular analyses that had previously been limited to microbes. Like the methods themselves, the history of recombinant DNA technology and DNA sequencing is dense, featuring a large cast of characters, including Paul Berg, Stanley Cohen, Herbert Boyer, Walter Gilbert, Frederick Sanger, and many others. Readers interested in learning more about the stories behind this vast topic may be interested in Life Sciences Foundation, "The Invention of Recombinant DNA Technology," Medium, November 11, 2015, https://medium.com/lsf-magazine/the-invention-of-recombinant-dna-technology-e040a8a1fa22.

140 **they line up:** Homeotic mutants were first observed early in the twentieth century, but the basis for their existence was a mystery. Ed Lewis, who developed the EMS mutagenesis technique used in the Heidelberg screen, showed that a family of homeobox-containing transcription factors likely arose from a single ancestral gene earlier in evolution, the duplicated genes forming an ordered cluster whose chromosomal arrangement specifies the identity of each segment along the fly's

body. (Hence, the four-winged *bithorax* fly results from a misspecification of one of the fly's body segments, converting a wingless thoracic segment into a wing-bearing one.)

141 **the genes that cause human diseases:** Lawrence Reiter, Lorrain Potocki, Sam Chien, Michael Gribskov, and Ethan Bier, "A Systematic Analysis of Human Disease-Associated Gene Sequences in *Drosophila melanogaster*," *Genome Research* 11, no. 6 (June 2001): 1114–25.

142 **"non-coding" RNA products:** In addition to the transcriptional repressors and activators introduced in Chapter 4, which control the expression of a gene by regulating the rate with which its mRNA message is transcribed, an analysis of worm mutants revealed an entirely new mechanism of gene regulation. These genes encode RNA products—a class of molecules called "micro-RNAs"—that find and degrade mRNA molecules with complementary nucleotide sequences, thus turning genes OFF by eliminating the messenger. The science of gene silencing by non-coding RNAs, or "RNA interference," has become its own major discipline.

143 **biologists remain far from achieving:** For more than a decade, biologists have worked, with some success, to create synthetic life-forms—microbes that could partially satisfy Feynman's challenge. Creating a synthetic animal (whether this is wise or not) is beyond the reach of current technology but *modeling* animal development is not. *C. elegans*'s simplicity has again made it the poster child for such attempts, and several groups have tried to model all elements of worm biology. One of the largest efforts is an international consortium, called OpenWorm, which seeks to fully simulate the worm's brain, body, and behavior; see their YouTube video *OpenWorm Open House: Introduction to OpenWorm Foundation*, https://www.youtube.com/watch?v=ROoZHLemRAs.

143 **"The frog remains":** ("Le biologist passe, la grenouille reste.") Jean Rostand, *Inquiétudes d'un biologiste* (Paris: Stock, 1967).

CHAPTER 6: DIRECTIONS, PLEASE!

147 **In mammals:** Although gastrulation is a universal feature of development in all animals, the details vary dramatically from species to species. The rate of cell cleavage is slower in mammals compared to other organisms, and the directionality of the first divisions—whether cells divide in a plane or at 90-degree angles—also differs across the animal kingdom. The biggest difference is location. Whereas most of the species we have considered so far develop outside the mother's body, mammalian development takes place in a uterus. In mice and humans, gastrulation takes place within a week of fertilization, while in other species, like cows, it can occur weeks later.

148 **three embryonic *germ layers*:** As the epiblast gives rise to the three germ layers and their descendants, a similar process of differentiation plays out in the placenta, where trophectoderm cells differentiate into the several lineages that will help the nascent embryo insert itself into the uterine wall and integrate with the maternal circulation. In many ways, the placenta is a sacrificial twin of the embryo, a complex tissue lacking a postnatal future of its own.

149 **the outside world:** Although it is intuitively accurate to think of the intestines as being "inside" the body, the inner surface of the gut tube—the "lumen"—is technically "outside." The entirety of the gut lumen, from mouth to anus, is enclosed

by a layer of intestinal epithelial cells that keeps its contents—food, waste, and bacteria—from entering the body. We can think of this as the equivalent of passing through an international terminal without clearing border control. Other "endoderm-derived" organs, including the lung, liver, and pancreas, have their own systems of tubes that connect to the gut tube at different points. These tubes are lined by epithelial cells that are continuous with the gut epithelium, and thus their lumens can also be thought of as connected to the "outside" world.

150 **the body's customs officers:** An example of how precisely the intestinal epithelium regulates what can enter the body, and what remains excluded, is the common condition of lactose intolerance. The disaccharide lactose (the same sugar Monod used in his studies of bacterial diauxie) is broken into its two constituent sugar subunits—glucose and galactose—by the enzyme lactase. When this enzyme is absent, or present at low levels, the bond between glucose and galactose is not broken. Because lactose itself cannot be absorbed (as the epithelial transport system is designed solely for its subunits), the undigested sugar travels through the entirety of the gut tube, carrying with it other molecules that lead to indigestion and diarrhea.

151 **a dozen or so distinct systems:** The most prevalent and evolutionarily conserved signals acting to shape the embryo fall into the following families: the Bone Morphogenetic Protein or Transforming Growth Factor family, the Fibroblast Growth Factor pathway, the Notch pathway, the Wingless/Int-1 pathway, and the Hedgehog pathway.

153 **vital for the placenta:** Placentation is a unique feature of mammals. Other branches of the animal kingdom—insects, amphibians, birds, fish, etc.—lack a placenta and therefore have no need for trophectoderm. While we tend to give the placenta little regard, it is a remarkable organ, and its development parallels that of the "embryo proper." Specifically, the placenta develops through a set of comparable cell-fate decisions and morphogenetic events that allow it to interface with the maternal blood supply inside the uterus. Consequently, placental defects remain one of the most common causes of stillbirth and miscarriage.

153 **the "inside-outside" model:** Andrzej Tarkowski and Joanna Wróblewska, "Development of Blastomeres of Mouse Eggs Isolated at the 4- and 8-Cell Stage," *Journal of Embryology and Experimental Morphology* 18, no. 1 (August 1967): 155–80.

153 **the fate of individual morula cells:** H. Balakier and R. A. Pedersen, "Allocation of Cells to Inner Cell Mass and Trophectoderm Lineages in Preimplantation Mouse Embryos," *Developmental Biology* 90, no. 2 (April 1982): 352–62.

155 **point the same way:** Planar cell polarity has been best studied in the fly, where mutations in certain proteins cause the fly's hairs to adopt a swirled rather than straight pattern. Examples include the *Drosophila* Van Gogh (*vang*) and Starry Night (*stan*) mutations, named as an homage to the Dutch artist's most famous painting.

158 ***convergent extension*:** Much of the molecular biology and cellular mechanics of this morphogenetic process were worked out by Ray Keller, a biologist at the University of California, Berkeley. See Ray Keller and Ann Sutherland, "Convergent Extension in the Amphibian, *Xenopus laevis*," *Current Topics in Developmental Biology* 136 (2020): 271–317.

159 **These forces:** Nandan Nerurkar, ChangHee Lee, L. Mahadevan, and Clifford Tabin, "Molecular Control of Macroscopic Forces Drives Formation of the Vertebrate Hindgut," *Nature* 565 (January 2019): 480–84; Amy Shyer, Tyler Huycke, Chang-

Hee Lee, L. Mahadeva, and Clifford Tabin, "Bending Gradients: How the Intestinal Stem Cell Gets Its Home," *Cell* 161, no. 3 (April 2015): 569–80.

161 **an organization to the cellular mass:** Philip Townes and Johannes Holtfreter, "Directed Movements and Selective Adhesion of Embryonic Amphibian Cells," *Journal of Experimental Zoology* 128, no. 1 (1955): 53–120.

163 **a badminton court:** Ewald Weibel, "What Makes a Good Lung?," *Swiss Medical Weekly* 139, no. 27–28 (July 2009): 375–86.

163 **Nature employs other methods:** Other tube-forming mechanisms include cavitation, akin to the "hollowing out" of rock that occurs during the formation of limestone caves. (Such hollowing can even occur inside a single cell, where in tissues like the *Drosophila* trachea it leads to "seamless" tubes.) Tubes can also form through a process of differential adhesion, or through the coalescing of small fluid-filled "microlumens" to form a continuous, open conduit. For reviews on tube formation, see Brigid Hogan and Peter Kolodziej, "Organogenesis: Molecular Mechanisms of Tubulogenesis," *Nature Reviews Genetics* 3, no. 7 (July 2002): 513–23; Luisa Iruela-Arispe and Greg Beitel, "Tubulogenesis," *Development* 140, no. 14 (July 2013): 2851–55; Ke Xu and Ondine Cleaver, "Tubulogenesis during Blood Vessel Formation," *Seminars in Cell and Developmental Biology* 22, no. 9 (December 2011): 993–1004.

164 **one of nature's best-kept secrets:** Ian Conlon and Martin Raff, "Size Control in Animal Development," *Cell* 96, no. 2 (January 1999): 235–44; Alfredo Penzo-Mendez and Ben Stanger, "Organ Size Regulation in Mammals," *Cold Spring Harbor Perspectives in Biology* 7, no. 9 (July 2015): a019240.

164 **the diminutive stature of small breeds:** Nathan Sutter, Carlos Bustamante, Kevin Chase, et al., "A Single IGF1 Allele is a Major Determinant of Small Size in Dogs," *Science* 316 (April 2007): 112–15.

164 **"as when we speak of a small elephant":** Darcy Thompson, *On Growth and Form* (Cambridge: Cambridge University Press, 1942), 24.

165 **the grafted cells "knew":** Ross Harrison, "Some Unexpected Results of the Heteroplastic Transplantation of Limbs," *Proceedings of the National Academy of Sciences USA* 10, no. 2 (February 1924): 69–74; Victor Twitty and Joseph Schwind, "The Growth of Eyes and Limbs Transplanted Heteroplastically between Two Species of Amblystoma," *Journal of Experimental Zoology* 59, no. 1 (February 1931): 61–86.

166 **it remained small:** Ben Stanger, Akemi Tanaka, and Douglas Melton, "Organ Size Is Limited by the Number of Embryonic Progenitor Cells in the Pancreas but Not the Liver," *Nature* 445 (February 2007): 886–91.

167 **amounts of energy:** Recently, researcher Yuval Dor at Hebrew University reported an unexpected correlation between life span and the size of a particular type of cell—the pancreatic acinar cell. Responsible for making the enzymes for digesting food, pancreatic acinar cells are among the largest cells in the body. Surprisingly, these cells vary in size nearly tenfold across different species in a paradoxical fashion: the largest animals have the smallest acinar cells. This relationship between acinar cell size and longevity—smaller cells correlate with longer life span—is even stronger than the relationship between body size and life span. See Shira Anzi, Miri Stolovich-Rain, Agnes Klochendler, et al., "Postnatal Exocrine Pancreas Growth by Cellular Hypertrophy Correlates with a Shorter Lifespan in Mammals," *Developmental Cell* 45, no. 6 (June 2018): 726–37.

167 **of limited use in understanding size:** To understand why the genetic approach lacks the same power to resolve questions of size control that it has regarding ques-

tions of embryonic patterning or cell-fate determination, consider how a genetic screen might be designed to study the problem. Such a screen would likely look for mutants that result in stunted growth or overgrowth, the phenotypes that should result if the genes controlling organ size are disrupted. But this would merely reveal genes constituting the *machinery* of growth. The underlying control mechanism—how a tissue *senses* how big it needs to be or whether it has reached the proper size—would be much harder to identify against this background.

169 **strong predictor of a cell's later fate:** Hui Yi Grace Lim, Yanina Alvarez, Maxime Gasnier, et al., "Keratins Are Asymmetrically Inherited Fate Determinants in the Mammalian Embryo," *Nature* 585 (September 2020): 404–9.

169 **in a matter of days:** John Murray and Zhirong Bao, "Automated Lineage and Expression Profiling in Live *Caenorhabditis elegans* Embryos," *Cold Spring Harbor Protocols* 8 (August 2012): pdb.prot070615.

INTERMEZZO: MATURATION

177 **In a recent essay:** Scott Gilbert, "Developmental Biology, the Stem Cell of Biological Disciplines," *PLoS Biology* 15, no. 12 (December 2017): e2003691. In addition to authoring the field's definitive textbook, the aptly named *Developmental Biology*, Gilbert is also a master historian of science.

177 **all cells come from other cells:** Robert Remak, "Uber die embryologische Grundlage der Zellenlehre," *Archiv für Anatomie, Physiologie und Wissenschaftliche Medicin* (1862): 230–41.

177 **"by the same law":** Rudolf Virchow, *Cellular Pathology as Based upon Physiological and Pathological Histology*, trans. Frank Chance (London: John Churchill, 1859).

CHAPTER 7: STEM CELLS

180 **McCulloch the dreamer and Till the go-getter:** Joe Sornberger, *Dreams and Diligence* (Toronto: University of Toronto Press, 2011), 30.

181 **to crawl out of the sea:** For a highly readable account of the first animals to begin making land their home, see Neil Shubin, *Your Inner Fish* (New York: Vintage Books, 2008).

183 **uranium's radioactive properties:** Becquerel's discovery of radioactivity was an accident. He found that placing uranium next to a photographic plate in the dark could expose the film, creating images. He believed that uranium's ability to make these images was due to a hypothetical ability to absorb, and later emit, rays of light from the sun. Becquerel chose to test this idea by showing that sun-exposed uranium could later expose photographic plates that were kept in the dark. But on the day that he set out to run his experiment, his hometown of Paris was overcast, preventing him from letting the uranium absorb the supposed sun-derived rays. Becquerel chose to expose the film anyway, expecting to see nothing, but instead observed a bright image in the pattern of the uranium crystals—proof that the uranium had emitted film-exposing rays on its own. It has subsequently come to light that French photographer Claude Félix Abel Niépce de Saint-Victor had made a similar observation nearly four decades earlier. Thus, credit for the discovery of radioactivity more appropriately belongs to Niépce de Saint-Victor.

183 **This man-made form of cobalt:** Cobalt-60 does not occur naturally. Cobalt's atomic

weight, in its natural form, is 59. Consequently, the radioactive isotope—cobalt-60—is created through high-energy physics, by bombarding naturally occurring cobalt-59 with neutrons.

187 **between four and nine "gray":** The gray (abbreviated Gy) is a unit for measuring dose of radiation *absorbed* by a tissue and is thus measured in terms of the amount of energy delivered (in joules) divided by the mass of tissue exposed. The formal definition of a gray is one joule of radiation energy delivered to one kilogram of matter.

193 **"ontogeny recapitulates phylogeny":** Haeckel believed that during development (ontogeny), an embryo must pass through all the evolutionary steps (phylogeny) that preceded its emergence. In some cases, evolutionary precursors can be observed in embryos—such as gill-like structures and webbed feet of mammalian embryos—and the embryos of many different animal species resemble one another until late in development. But for the most part, embryos do not retrace the steps of their evolutionary forebears.

193 **"*stammzellen*":** Miguel Ramalho-Santos and Holger Willenbring, "On the Origin of the Term 'Stem Cell,'" *Cell Stem Cell* 1, no. 1:35–38.

196 **Till and McCulloch's first papers:** E. A. McCulloch and J. E. Till, "The Radiation Sensitivity of Normal Mouse Bone Marrow Cells, Determined by Quantitative Marrow Transplantation into Irradiated Mice," *Radiation Research* 13 (1960): 115–25; J. E. Till and E. A. McCulloch, "A Direct Measurement of the Radiation Sensitivity of Normal Mouse Bone Marrow Cells," *Radiation Research* 14 (1961): 213–22.

196 **a birth announcement of sorts:** A. J. Becker, E. A. McCulloch, and J. E. Till, "Cytological Demonstration of the Clonal Nature of Spleen Colonies Derived from Transplanted Mouse Marrow Cells," *Nature* 197 (February 1963): 452–54.

197 **Each time a stem cell divides:** Becker's experiment provided evidence of both features. The fact that each splenic colony contained a diversity of blood lineages (red blood cells, white blood cells, etc.) confirmed that it had arisen from a cell that was multipotent. Furthermore, the property of self-renewal was implicit in the size of the colony, which contained millions of cells. A year later, OCI scientist Louis Siminovitch showed that cells isolated from the nodule of one mouse could give rise to lumpy spleens in a second mouse—a process known as "serial transplantation." It was formal proof that a colony, the product of a single cell, still contained cells capable of repeating the process.

197 **"A million dollars!":** Ann Parson, *The Proteus Effect* (Washington, DC: Joseph Henry Press, 2004), 61.

197 **"Going from one cell to a million":** Parson, *Proteus Effect*, 61.

199 **more than 1.5 million transplants:** Dieter Niederwieser, Helen Baldomero, Yoshiko Atsuta, et al., "One and a Half Million Hematopoietic Stem Cell Transplants (HSCT)," *Blood* 134, no. S1 (November 2019): 2035.

CHAPTER 8: CELLULAR ALCHEMY

202 **This proofreading exercise:** Most animals produce offspring via sexual reproduction, whereby gametes that are *haploid* (i.e., have one set of chromosomes) give rise to cells that are diploid (i.e., have two sets of chromosomes). Sexual reproduction is thought to have evolved for its ability to enhance variation within a population, making it more suited to adapt to selective pressures. But a side benefit of a diploid

genome is that when one copy gets damaged, there is always a second copy around with which to repair it—via homologous recombination.

203 **1 out of 1,000 injected cells:** Kirk Thomas, Kim Folger, and Mario Capecchi, "High Frequency Targeting of Genes to Specific Sites in the Mammalian Genome," *Cell* 44 (February 1986): 419–28.

203 **Capecchi was not alone:** Oliver Smithies, Ronald Gregg, Sallie Boggs, Michael Koralewski, and Raju Kucherlapati, "Insertion of DNA Sequences into the Human Chromosomal β-Globin Locus by Homologous Recombination," *Nature* 317 (September 1985): 230–34.

204 **useful for immunological studies:** The development of inbred strains of mice and other breeding techniques was critical for determining the details of immune compatibility, features that are now used to match organ donors with recipients prior to transplantation.

204 **the tumors were transplantable:** Leroy Stevens, "Studies on Transplantable Testicular Teratomas of Strain 129 Mice," *Journal of the National Cancer Institute* 20, no. 6 (June 1958): 1257–75.

205 **Again, teratomas grew:** Davor Solder, Nikola Skreb, and Ivan Damjanov, "Extrauterine Growth of Mouse Egg-Cylinders Results in Malignant Teratoma," *Nature* 227 (August 1970): 503–4.

206 **an individual teratoma cell:** Lewis Kleinsmith and G. Barry Pierce, "Multipotentiality of Single Embryonal Carcinoma Cells," *Cancer Research* 24 (October 1964): 1544–51.

206 **strains with different coat colors:** Laila Moustafa and Ralph Brinster, "Induced Chimaerism by Transplanting Embryonic Cells into Mouse Blastocysts," *Journal of Experimental Zoology* 181, no. 2 (August 1972): 193–201.

206 **called *chimeric mice*:** The first chimeric mice were created by Polish embryologist Andrzej Tarkowski and American embryologist Beatrice Mintz in the 1960s by fusing together early embryos prior to the morula stage. The advantage of Brinster's approach was that it simplified the formation of chimeras, making it possible to incorporate other types of cells into a developing embryo. Technically speaking, each chimera may have up to four parents, as the donor and host each have two parents of their own.

208 **patches of dark hair:** Ralph Brinster, "The Effect of Cells Transferred into the Mouse Blastocyst on Subsequent Development," *Journal of Experimental Medicine* 140, no. 4 (October 1974): 1049–56. Although only one mouse showed clear evidence of chimerism in this initial experiment, other mice exhibited indirect evidence of EC cell contributions.

209 **give rise to teratomas:** Leroy Stevens, "The Development of Teratomas from Intratesticular Grafts of Tubal Mouse Eggs," *Journal of Embryology and Experimental Morphology* 20, no. 3 (November 1968): 329–41.

210 **they were sterile:** Tumor cells carry mutations and chromosomal alterations. The fact that EC cells came from teratomas had been useful for their derivation (a result of the rapid adaptability of tumor cells to culture). But when it came to "germline transmission"—the ability to foster a new generation—this asset became a liability. While most differentiated cells can tolerate the genetic abnormalities of the EC cells, gametes and/or zygotes cannot.

210 **months of trial and error:** M. J. Evans and M. H. Kaufman, "Establishment in Culture of Pluripotential Cells from Mouse Embryos," *Nature* 292 (July 1981): 154–56.

210 **in her own laboratory:** Gail Martin, "Isolation of a Pluripotent Cell Line from Early Mouse Embryos Cultured in Medium Conditioned by Teratocarcinoma Stem Cells," *Proceedings of the National Academy of Sciences USA* 78, no. 12 (December 1981): 7634–38.

213 **replacing the functional copy:** Kirk Thomas and Mario Capecchi, "Site-Directed Mutagenesis by Gene Targeting in Mouse Embryo-Derived Stem Cells," *Cell* 51, no. 3 (November 1987): 503–12.

213 **it could also fix them:** Thomas Doetschman, Ronald Gregg, Nobuyo Maeda, et al., "Targeted Correction of a Mutant HPRT Gene in Mouse Embryonic Stem Cells," *Nature* 33 (December 1987): 576–78.

213 **to "go germline":** Beverly Koller, Lora Hagemann, Thomas Doetschman, et al., "Germ-Line Transmission of a Planned Alteration Made in a Hypoxanthine Phosphoribosyltransferase Gene by Homologous Recombination in Embryonic Stem Cells," *Proceedings of the National Academy of Sciences USA* 86, no. 22 (November 1989): 8927–31.

214 **to make mutant mice:** Suzanne Mansour, Kirk Thomas, and Mario Capecchi, "Disruption of the Proto-oncogene *int-2* in Mouse Embryo-Derived Stem Cells: A General Strategy for Targeting Mutations to Non-selectable Genes," *Nature* 336 (November 1988): 348–52.

214 **The gene they chose:** Kirk Thomas and Mario Capecchi, "Targeted Disruption of the Murine *int-1* Proto-oncogene Resulting in Severe Abnormalities in Midbrain and Cerebellar Development," *Nature* 346 (August 1990): 847–50.

215 **resolved the technical issues:** James Thomson, Joseph Itskovitz-Eldor, Sander Shapiro, et al., "Embryonic Stem Cell Lines Derived from Human Blastocysts," *Science* 282 (November 1998): 1145; Michael Shamblott, Joyce Axelman, Shuping Wang, et al., "Derivation of Pluripotent Stem Cells from Cultured Human Primordial Germ Cells," *Proceedings of the National Academy of Sciences USA* 95, no. 23 (November 1998): 13726–31.

216 **and other tissues:** This technique, known as a "conditional" knockout, turned out to be another important advance. Recall that the phenomenon of "embryonic lethality"—the feature that helped Wieschaus and Nüsslein-Volhard narrow down their candidates in the Heidelberg screen (Chapter 5)—precludes animals with a knockout of an embryonic-lethal gene from reaching adulthood. Consequently, the early days of gene knockout technology were filled with reports of gene functions during embryogenesis. As researchers became more interested in the function of genes in adult tissues, conditional-knockout approaches became critical, as they allowed scientists to bypass the pothole of embryonic lethality.

216 **the products of ESC-differentiation:** The biggest difference between the cellular products of ESC differentiation in vitro and their normal counterparts in vivo appears to be a failure in "maturation," the cellular transition from an embryonic to an adult functional state. The basis of defective maturation is unclear, but it may be related to the microenvironments or cellular neighborhoods that exist inside the body but are hard to reproduce in a dish.

216 **"the central predicament of his young presidency":** Richard Lacayo, "How Bush Got There," *Time*, August 20, 2001.

217 **"morally unacceptable":** Joseph Fiorenza, "Response to the Bush Policy from the U.S. Conference of Catholic Bishops," Catholic Culture, https://www.catholicculture.org/culture/library/view.cfm?recnum=3960.

219 **the pluripotent state was "dominant":** Masako Tada, Yousuke Takahama, Kuniya Abe, Norio Nakatsuji, and Takashi Tada, "Nuclear Reprogramming of Somatic Cells by in Vitro Hybridization with ES Cells," *Current Biology* 11, no. 19 (October 2001): 1553–58.

220 **convert a fibroblast into a myoblast:** Robert Davis, Harold Weintraub, and Andrew Lassar, "Expression of a Single Transfected cDNA Converts Fibroblasts to Myoblasts," *Cell* 51, no. 6 (December 1987): 987–1000.

221 **the "Yamanaka factors":** The four factors were Oct4, Klf4, Sox2, and c-Myc. Kazutoshi Takahashi and Shinya Yamanaka, "Induction of Pluripotent Stem Cells from Mouse Embryonic and Adult Fibroblasts Cultures by Defined Factors," *Cell* 126, no. 4 (August 2006): 663–76.

222 ***induced pluripotent stem cells* (iPSCs):** The "Yamanaka factors" are not the only genes that can kick-start the reprogramming process; several other genes, or chemicals, can substitute for each component. The features that allow these four genes and their stand-ins to turn back the developmental clock so thoroughly remain areas of active investigation. But among the essential steps are alterations in the epigenome—the chemical modifications to DNA and its associated proteins (*chromatin*) that cause a wholesale activation of genes associated with stem cell identity and silencing of genes associated with fibroblast identity. These mechanisms are described in greater detail in Chapter 11.

222 **difficult to find significant differences:** Jiho Choi, Soohyun Lee, William Mallard, et al., "A Comparison of Genetically Matched Cell Lines Reveals the Equivalence of Human iPSCs and ESCs," *Nature Biotechnology* 33, no. 11 (November 2015): 1173–81.

224 **has led to a clinical trial:** Brian Wainger, Eric Macklin, Steve Vucic, et al., "Effect of Ezogabine on Cortical and Spinal Motor Neuron Excitability in Amyotrophic Lateral Sclerosis: A Randomized Clinical Trial," *JAMA Neurology* 78, no. 2 (February 2021): 186–96.

225 **became apparent only with iPSC:** Max Cayo, Sunil Mallanna, Francesca Di Furio, et al., "A Drug Screen Using Human iPSC-Derived Hepatocyte-Like Cells Reveals Cardiac Glycosides as a Potential Treatment for Hypercholesterolemia," *Cell Stem Cell* 20, no. 4 (April 2017): 478–89.

225 **hepatocytes can come from a variety of individuals:** Robert Schwartz, Kartik Trehan, Linda Andrus, et al., "Modeling Hepatitis C Virus Infection Using Human Induced Pluripotent Stem Cells," *Proceedings of the National Academy of Sciences USA* 109, no. 7 (February 2012): 2544–48.

CHAPTER 9: ONE CELL RUN AMOK

227 **"Cancer didn't bring me to my knees":** As quoted in Lynn Elber, "Hanks, Roberts among Stars on 'Stand Up to Cancer,'" *Spokesman-Review*, September 8, 2012.

227 **tumors and embryos have more in common:** Nicole Aiello and Ben Stanger, "Echoes of the Embryo: Using the Developmental Biology Toolkit to Study Cancer," *Disease Models and Mechanisms* 9, no. 2 (February 2016): 105–14.

228 **Boveri codified his ideas:** Theodor Boveri, "Concerning the Origin of Malignant Tumours (1914)," trans. and annotated by Henry Harris, *Journal of Cell Science* 121, no. S1 (January 2008): 1–84.

229 **This regulated and dynamic balance:** We know a great deal about the pathways

that control growth at the cellular level. But as we saw earlier (Chapter 6), we do not understand how these growth-promoting and growth-suppressing signals control size at the level of the entire animal, the unsolved problem of size control.

229 **mutations abolish this balance:** In this context, the word "mutations" can take on many different meanings. In the simplest sense, cancer-causing mutations involve single-nucleotide alterations in a gene's DNA sequence, which may alter the corresponding amino acid sequence to augment the cancer-promoting activity of an oncogene or abrogate the tumor-inhibiting activity of a tumor suppressor gene. On a larger molecular scale, mutations may involve the kinds of chromosomal alterations Boveri envisioned—"deletions" or "amplifications" of large segments of a chromosome, or "translocations" that fuse together two distinct chromosomes. Finally, the expression of an oncogene or tumor suppressor gene can also be modulated through *epigenetic* modifications—heritable alterations in gene expression (Chapter 11).

229 **tissue maintenance is the order of the day:** From our earlier discussion of stem cells, we saw that tissues can maintain themselves either from stem cells (in tissues with rapid turnover, like the intestine, skin, or blood) or from existing cells (in tissues with slow turnover, where existing cells divide to replace dying cells). In the former, oncogenes exert their normal function, driving the rapid proliferation of stem cell progeny to fill the substantial need for new cells, while in the latter, tumor suppressor genes keep cells in their quiescent state.

230 **a tumor's parentage:** The recognition that a tumor's origins cannot be determined based on its appearance under the microscope is a relatively recent development in cancer biology. For example, the formal name given to pancreatic cancer is "pancreatic ductal adenocarcinoma," implying that the tumors have their origins in the ducts of the pancreas. But as we discuss in this section, this may or may not be true for a given tumor. In general, carcinomas resemble one another regardless of the tissue of origin, leading to the clinical scenario of "metastatic carcinoma of unknown primary"—cases in which metastatic cancer is present but no primary tissue of origin can be identified.

231 **genetically engineered mouse models:** Douglas Hanahan, Erwin Wagner, and Richard Palmiter, "The Origins of Oncomice: A History of the First Transgenic Mice Genetically Engineered to Develop Cancer," *Genes and Development* 21, no. 18 (September 2007): 2258–70. The traditional method for evaluating whether a drug possesses antitumor effects in vivo is *xenotransplantation*, in which tumor cells from a cultured cell line are engrafted into the flank of a mouse and the effect of one or more drugs is assessed. Xenotransplantation requires the use of immunodeficient mice (so that the human tumor cells are not rejected by the mouse's immune system) and does not allow the tumor to go through its normal evolutionary process (implanted tumor cells are already at an advanced stage at the time of engraftment). Genetically engineered mouse models (GEMMs) circumvent these limitations because the tumors evolve spontaneously over time as they grow in the mouse, and they coexist with an intact immune system. The main limitation of GEMMs, of course, is that they give rise to mouse tumors, not human tumors.

231 **The model was created:** Sunil Hingorani, Lifu Wang, Asha Multani, et al., "*Trp53R172H* and *KrasG12D* Cooperate to Promote Chromosomal Instability and Widely Metastatic Pancreatic Ductal Adenocarcinoma in Mice," *Cancer Cell* 7, no. 5 (May 2005): 469–83.

232 **lost the ability to divide:** Brain tumors, which afflict nearly 1 in 10,000 people each year, may appear to contradict this claim, but they do not. Brain tumors, as a rule, do not arise from neurons. Instead, it is the supporting cells of the central nervous system—the glial cells and astrocytes, which *do* divide—that are the likely source of most brain tumors.

232 **the most common and lethal malignancies:** Currently, the five most lethal human tumors in the United States are (in order) carcinomas of the lung, colon and rectum, pancreas, breast, and prostate, which according to the American Cancer Society are responsible for more than half of the 600,000 deaths due to all cancers combined. (NB: These numbers omit two kinds of skin cancer—squamous cell carcinoma and basal cell carcinoma—that are among the least lethal tumor types.) The reasons for the high prevalence and lethality of carcinomas are not known, but one possibility is that epithelia are more exposed to the "outside world," making them subject to greater injury and thus at greater risk of suffering mutations. This idea is supported by the fact that carcinomas of the skin are the most common types of cancer, the result of sun-induced mutations. However, other observations make the injury model unsatisfying. Moreover, the malignant spectrum differs in other species. Mice in captivity, for example, commonly die of leukemias, lymphomas, and sarcomas, tissues that are less exposed to external elements.

233 **mutations affecting the ABL oncogene:** Mutations causing the ABL oncogene to become hyperactivated are commonly associated with an error causing chromosomes 9 and 22 to become fused together. The discovery in 1960 of this unique abnormality—dubbed the "Philadelphia chromosome" after the site of its discovery—provided the first link between cancer and a specific genetic event.

234 **The most important factor:** Katherine Hoadley, Christina Yau, Toshinori Hinoue, et al., "Cell-of-Origin Patterns Dominate the Molecular Classification of 10,000 Tumors from 33 Types of Cancer," *Cell* 173, no. 2 (April 2018): 291–304.

235 **There is evidence on both sides:** Cancer-associated de-differentiation, while common, is not universal. In so-called neuroendocrine tumors, for example, cancer cells continue to produce hormones like insulin, a highly specialized duty. While these tumor cells can be distinguished from normal insulin-producing cells, they still retain this highly differentiated feature.

236 **markers bearing the names:** One of the great advantages of working in the blood system is that cells can be distinguished based on the presence of distinctive proteins on the cell surface, so-called "cluster of differentiation" (CD) markers. By staining cells with a mixture of antibodies recognizing different CD proteins, it is possible to recover cells that share the same cell-surface phenotype. There are well over 200 different CD proteins, many of which are known by other names that better reflect the protein's biological functions, but the CD nomenclature provides scientists with a common vocabulary for describing different types of cells.

239 **each of these steps occurs infrequently:** Patients with metastatic cancers, even aggressive ones, seldom have more than one to two dozen detectable metastatic lesions (although on occasion, patients will present with hundreds of metastases). Nevertheless, these numbers reflect only those metastases that can be detected by standard imaging—CT scans or MRIs. But for many or most patients with metastasis, this is just the tip of the iceberg, as many more metastases—so called *micrometastases*—may exist below the limit of detection. Nevertheless, even

including these microscopic lesions, the burden of metastasis is far smaller than one might expect based on first principles.

241 **cancers must recruit new blood vessels:** Judah Folkman, "Tumor Angiogenesis: Therapeutic Implications," *New England Journal of Medicine* 285, no. 21 (November 1971): 1182–86.

243 **tumor-promoting and tumor-suppressing flavors:** Andrew Rhim, Paul Oberstein, Dafydd Thomas, et al., "Stromal Elements Act to Restrain, Rather Than Support, Pancreatic Ductal Adenocarcinoma," *Cancer Cell* 25, no. 6 (June 2014): 735–47; Berna Ozdemir, Tsvetelina Pentcheva-Hoang, Julienne Carstens, et al., "Depletion of Carcinoma-Associated Fibroblasts and Fibrosis Induces Immunosuppression and Accelerates Pancreas Cancer with Reduced Survival," *Cancer Cell* 25, no. 6 (June 2014): 719–34; Erik Sahai, Igor Astsaturov, Edna Cukierman, et al., "A Framework for Advancing Our Understanding of Cancer-Associated Fibroblasts," *Nature Reviews Cancer* 20, no. 3 (March 2020): 174–86.

CHAPTER 10: EYE OF NEWT AND TOE OF FROG

245 **the accident that would leave her paralyzed:** Conversation with the author, September 3, 2009.

246 **300,000 people in the United States:** National Spinal Cord Injury Statistical Center, *Facts and Figures at a Glance* (Birmingham, AL: University of Alabama at Birmingham, 2021).

246 **An analysis published in 1998:** Monroe Berkowitz, Paul O'Leary, Douglas Kruse, and Carol Harvey, *Spinal Cord Injury: An Analysis of Medical and Social Costs* (New York: Demos Medical Publishing, 1998).

246 **Bowel and bladder dysfunction:** Continence is maintained by sphincters, which remain closed until the brain sends a neuronal signal (through the spinal cord) telling the sphincters to relax. This circuit is disrupted with spinal cord injury, thus causing the sphincters to remain constricted. Prior to World War II, the leading cause of death from SCI was kidney failure, the result of urinary retention and backflow into the kidney. The subsequent introduction of urinary catheterization—which bypasses the sphincter—has all but eradicated this problem.

247 **all the neurons we will ever have:** Neurogenesis—the birth of new neurons—is most robust during the second trimester. Paradoxically, neuronal precursor cells divide so rigorously during this period that the embryo creates more neurons than it ultimately needs. The excess—perhaps more than half of the neurons formed in the first place—are then culled in the weeks or months surrounding birth via programmed cell death (Chapter 5) through a process known as "pruning." Consequently, while the rest of the body continues accruing cells during the first decades of life, the nervous system remains static. While there is evidence that new neurons are produced in certain regions of the adult human brain, this occurs at such a low level that it is not thought to have a meaningful impact on learning or regeneration. (This is paradoxical, given that the mind is the body part with the greatest capacity for growth during adult life.) The acquisition of a new language or mastery of a video game does not entail the birth of new neurons, the way one might install a "memory card" into a personal computer when more RAM is needed. The reason for this near cessation of neuronal outgrowth is unknown—perhaps adding

new elements into the complex circuits of the brain would only slow things down. Whatever the reason, learning is mediated by changes in synapses rather than the proliferation of cells, with profound implications for neurodegenerative diseases and strokes.

248 **suicide accounts for up to 10 percent:** P. Kennedy and L. Garmon-Jones, "Self-Harm and Suicide before and after Spinal Cord Injury: A Systematic Review," *Spinal Cord* 55, no. 1 (January 2017): 2–7.

250 **Mammals do not engage:** Mammalian regeneration (compensatory growth) is an extension of the normal mechanisms by which tissues maintain themselves, using either stem cells or the replication of existing cells to fuel turnover. The intestine is a good example of this physiological process, as the entire epithelial layer is replaced in a matter of days, even under normal circumstances. If the epithelial layer is injured—the result of radiation or chemotherapy, for example—it will be replaced by the remaining stem cells that withstood the damage. Thomas Hunt Morgan, of the Fly Room, coined the term *epimorphic regeneration* (Morgan studied regeneration before turning his attention to flies). His intent was to distinguish the dramatic regenerative abilities of organisms like newts and salamanders—which create whole extremities anew—from the more pedestrian powers of mammals, which can only rearrange existing tissue. For further details and a taxonomy of regenerative processes, see Bruce Carlson's *Principles of Regenerative Biology* (Burlington, MA: Academic Press, 2007).

251 **a smaller-than-normal limb:** Dorothy Skinner and John Cook, "New Limbs from Old: Some Highlights in the History of Regeneration in Crustacea," chap. 3 in *A History of Regeneration Research*, ed. Charles Dinsmore (Cambridge: Cambridge University Press, 1991).

251 **To his amazement:** Limb regeneration in crustaceans is thought to be related to the physiological program of "self-amputation." When the claw of a crab or lobster becomes entrapped, and its life is endangered, the animal can dislocate a joint at the base of the limb, allowing it to pull its body away from the trapped extremity and escape. Skinner and Cook, "New Limbs from Old."

251 **decapitated snails could grow new heads:** For centuries, the head had been considered the seat of the soul. Spallanzani's findings created a conundrum for religious thinkers, who now had to explain how a displaced soul could be so easily replaced. All this happened as the guillotine was gaining popularity in France, adding fodder to the public's growing fascination with decapitation.

251 **"snails are the talk of the town":** Marguerite Carozzi, "Bonnet, Spallanzani, and Voltaire on Regeneration of Heads in Snails: A Continuation of the Spontaneous Generation Debate," *Gesnerus* 42, nos. 2–3 (November 1985): 265–88.

253 **causing them to de-differentiate:** Given this de-differentiation, it is reasonable to ask whether limb regeneration employs similar pathways to those employed during programming to pluripotency. One study found that while the cells of the blastema express some of the reprogramming factors capable of inducing pluripotency, the cells do not enter a pluripotent state. For further reading, see Bea Christen, Vanesa Robles, Marina Raya, Ida Paramonov, and Juan Carlos Izpisua Belmonte, "Regeneration and Reprogramming Compared," *BMC Biology* 8, no. 5 (January 2010).

253 **Wolpert's model:** Lewis Wolpert, "Positional Information and the Spatial Pattern of Cellular Differentiation," *Journal of Theoretical Biology* 25, no. 1 (October 1969): 1–47.

254 **While mammals lack:** There are a few exceptions to this general rule in which mammals (including humans) can generate blastema-like cellular aggregates from which newly formed structures emerge. For example, if a young child should lose a fingertip, a blastema-like structure forms on the stump, allowing the regrowth of an entirely new fingertip (this phenomenon depends on a portion of the nail bed being retained, and the ability to regenerate a fingertip is lost as children grow older). Blastema formation is essential, for if the blastema is prevented from forming—should the amputation stump be oversewn (i.e., "stitched up") by a well-meaning physician, for example—the amputated portion will not regrow.

256 **more than a million new livers:** Ken Overturf, Muhsen Al-Dhalimy, Ching-Nan Ou, Milton Finegold, and Markus Grompe, "Serial Transplantation Reveals the Stem-Cell-Like Regenerative Potential of Adult Mouse Hepatocytes," *American Journal of Pathology* 151, no. 5 (November 1997): 1273–80.

257 **the ability to regenerate disappeared:** Kostandin Pajcini, Stephane Corbel, Julien Sage, Jason Pomerantz, and Helen Blau, "Transient Inactivation of Rb and ARF Yields Regenerative Cells from Postmitotic Mammalian Muscle," *Cell Stem Cell* 7, no. 2 (August 2010): 198–213.

260 **sophisticated artificial hands and legs:** It is almost impossible to keep up with the rate of advances in bioengineering. As new generations of prostheses provide greater function—motor control and rudimentary sensation—the cost to produce them is declining. An example can be seen in Guoying Gu, Ningbin Zhang, Haipeng Xu, et al., "A Soft Neuroprosthetic Hand Providing Simultaneous Myoelectric Control and Tactile Feedback," *Nature Biomedical Engineering* 464 (August 2021).

260 **live a decade longer:** Ye Zhang, Ulf-G. Gerdtham, Helena Rydell, and Johan Jarl, "Quantifying the Treatment Effect of Kidney Transplantation Relative to Dialysis on Survival Time: New Results Based on Propensity Score Weighting and Longitudinal Observational Data from Sweden," *International Journal of Environmental Research and Public Health* 17, no. 19 (October 2020): 7318.

261 **methods for growing human cells improved:** Magdalena Jedrzejczak-Silicka, "History of Cell Culture," chap. 1 in *New Insights into Cell Culture Technology*, ed. Sivakumar Joghi Thatha Gowder (London: IntechOpen, 2017); Rebecca Skloot, *The Immortal Life of Henrietta Lacks* (New York: Crown Publishers, 2010).

262 **state governments:** California, Connecticut, Maryland, New York, Illinois, and New Jersey have all provided some level of state support for regenerative medicine or stem cell initiatives.

263 **improvement in locomotion and gait:** Hans Keirstead, Gabriel Nistor, Giovanna Bernal, et al., "Human Embryonic Stem Cell-Derived Oligodendrocyte Progenitor Cell Transplants Remyelinate and Restore Locomotion after Spinal Cord Injury," *Journal of Neuroscience* 25, no. 19 (May 2005): 4694–705. See also *Paralyzed Rat Walks Again with Human Embryonic Stem Cells*, video posted by chrisclub March 23, 2009, YouTube, http://www.youtube.com/watch?v=5x8e2qsAVGc&feature=related.

263 **redirecting an individual's T cells:** Gideon Gross, Tova Waks, and Zelig Eshhar, "Expression of Immunoglobulin-T-Cell Receptor Chimeric Molecules as Functional Receptors with Antibody-Type Specificity," *Proceedings of the National Academy of Sciences USA* 86, no. 24 (December 1989): 10024–28.

264 **Ludwig was cured:** Bill Ludwig died of COVID in 2021; at the time, he was cancer-free. Marie McCullough, "Bill Ludwig, Patient Who Helped Pioneer Cancer

Immunotherapy at Penn, Dies at 75 of COVID-19," *Philadelphia Inquirer*, February 17, 2021.

265 **limiting factor:** Multiple technical and logistical difficulties hamper the more widespread implementation of islet transplantation beyond the limitations constraining all organ transplantation. Few medical centers have the capacity to expeditiously perform the considerable processing needed to isolate islets suitable for transplantation. In addition, there is substantial islet attrition during processing, which often means that the islets isolated from a single donor may be insufficient. Under those circumstances, two donors must be identified at the same time for transplantation to be practical.

266 **Researchers instilled millions of cells:** As is common for phase 1 clinical trials—where the priority is to establish a drug's safety rather than its efficacy—Shelton received half the dose of cells that the investigators thought would be needed to influence his diabetes. If no severe side effects are observed in several patients, the next set of patients receive a higher dose of the therapy in what is known as "dose-escalation."

266 **Shelton no longer needed insulin:** Shelton's early response to the therapy was reported in the *New York Times* by Gina Kolata (November 27, 2021), with more recent results presented in a June 2022 press release from Vertex ("Vertex Presents New Data from VX-880 Phase 1/2 Clinical Trial at the American Diabetes Association 82nd Scientific Sessions"). At the time of this writing, Vertex has reported that at least one other patient has shown improvements in blood sugar following an infusion of half the target dose of the stem cell–derived target, and one patient has received the full dose of cells with no adverse effects. Approximately 17 patients will be enrolled in this initial phase 1/2 clinical study. While the source of the cells used to make Vertex's product has not been disclosed publicly, it is possible that one day these methods will rely on iPSCs instead of ESCs, so that patients may, in effect, receive their own cells during therapy.

267 **no front-runner therapies have emerged:** Kazuyoshi Yamazaki, Masahito Kawabori, Toshitaka Seki, and Kiyohiro Houkin, "Clinical Trials of Stem Cell Treatment for Spinal Cord Injury," *International Journal of Molecular Sciences* 21, no. 11 (June 2020): 3994.

CHAPTER 11: DAY SCIENCE AND NIGHT SCIENCE

269 **"day science" and "night science":** François Jacob, *The Statue Within*, trans. Franklin Philip (New York: Basic Books, 1988), 296.

270 **"the cloud":** Uri Alon, "How to Choose a Good Scientific Problem," *Molecular Cell* 35, no. 6 (September 2009): 726–28.

270 **"learning the rules":** William Kaelin, "Why We Can't Cure Cancer with a Moonshot," Opinions, *Washington Post*, February 11, 2020.

272 **pass those identities along:** There are, of course, exceptions to this rule of heritability. For example, plasticity enables cells to change their identities following injury or certain experimental stimuli. But this only occurs in the setting of some physiological upheaval; left unperturbed, cells retain their sense of identity.

273 **coined the term "epigenetics":** Conrad Waddington, "The Epigenotype (1942)," *Endeavor* 1:18–20, reprinted in *International Journal of Epidemiology* 41, no. 1 (February 2012): 10–13. Also, to avoid any confusion, it is worth distinguishing the term

"epigenetics" from "epigenesis," which we encountered in Chapter 1. The former refers to a mechanism for encoding heritable information, while the latter refers to the piecemeal assembly of a body, a concept first articulated by Aristotle that later came to be the major theory contradicting preformationism.

274 **which genes to express and which to muzzle:** Arthur Riggs, "X Inactivation, Differentiation, and DNA Methylation," *Cytogenetics and Cell Genetics* 14, no. 1 (1975): 9–25; R. Holliday and J. E. Pugh, "DNA Modification Mechanisms and Gene Activity during Development," *Science* 187 (January 1975): 226–32.

275 **repository of cellular memory:** Gary Felsenfeld, "A Brief History of Epigenetics," *Cold Spring Harbor Perspectives in Biology* 6, no. 1 (January 2014): a018200; Tally Naveh-Many and Howard Cedar, "Active Gene Sequences Are Undermethylated," *Proceedings of the National Academy of Sciences USA* 78, no. 7 (July 1981): 4246–50; Reuven Stein, Yosef Gruenbaum, Yaakov Pollack, Aharon Razin, and Howard Cedar, "Clonal Inheritance of the Pattern of DNA Methylation in Mouse Cells," *Proceedings of the National Academy of Sciences USA* 79, no. 1 (January 1982): 61–65; Adrian Bird, Mary Taggart, Marianne Frommer, Orlando Miller, and Donald Macleod, "A Fraction of the Mouse Genome That Is Derived from Islands of Nonmethylated CpG-Rich DNA," *Cell* 40, no. 1 (January 1985): 91–99.

275 **mammalian embryos "erase":** As with most paradigms in biology, there are exceptions, and the erasure of DNA methylation is an important one. While the vast majority of methylated cytosines and other epigenetic marks are removed early in development, some remain. The incomplete nature of erasure creates the opportunities for "transgenerational" inheritance, discussed later in the chapter.

276 **subunits, called *nucleosomes*:** The configuration of cellular DNA organized into the repeating subunits now known as nucleosomes was first observed by husband-and-wife team Don and Ada Olins using electron microscopy (Ada Olins and Donald Olins, "Spheroid Chromatin Units [v Bodies]," *Science* 183 [January 1974]: 330–32). Shortly afterward, biochemist Roger Kornberg (who would later win the Nobel Prize for different work) proposed what has become the standard nucleosome model (Roger Kornberg, "Chromatin Structure: A Repeating Unit of Histones and DNA," *Science* 184 [May 1974]: 868–71).

277 **yeast cell's ability to transcribe a gene:** Linda Durrin, Randall Mann, Paul Kayne, and Michael Grunstein, "Yeast Histone H4 N-Terminal Sequence Is Required for Promoter Activation in Vivo," *Cell* 65, no. 6 (June 1991): 1023–31.

277 **researchers David Allis and Stuart Schreiber:** Jack Taunton, Christian Hassig, and Stuart Schreiber, "A Mammalian Histone Deacetylase Related to the Yeast Transcriptional Regulator Rpd3p," *Science* 272 (April 1996): 408–11; James Brownell, Jianxin Zhou, Tamara Ranalli, et al., "Tetrahymena Histone Acetyltransferase A: A Homology to Yeast Gcn5p Linking Histone Acetylation to Gene Activation," *Cell* 84, no. 6 (March 1996): 843–51.

278 **the best analogy:** Nessa Carey, *The Epigenetics Revolution* (New York: Columbia University Press, 2012), 68–69.

278 **nicknamed the "histone code":** Thomas Jenuwein and David Allis, "Translating the Histone Code," *Science* 293 (August 2001): 1074–80.

281 **the tendency to inherit:** Hugh Morgan, Heidi Sutherland, David Martin, and Emma Whitelaw, "Epigenetic Inheritance at the Agouti Locus in the Mouse," *Nature Genetics* 23 (November 1999): 314–18.

281 **higher risk of metabolic conditions:** Gian-Paolo Ravelli, Zena Stein, and Mervyn Susser, "Obesity in Young Men after Famine Exposure in Utero and Early Infancy," *New England Journal of Medicine* 295 (August 1976): 349–53; Rebecca Painter, Tessa Roseboom, and Otto Bleker, "Prenatal Exposure to the Dutch Famine and Disease in Later Life: An Overview," *Reproductive Toxicology* 20, no. 3 (September–October 2005): 345–52.

281 **whose *mothers* were fetuses:** L. H. Lumey and Aryeh Stein, "Offspring Birth Weights after Maternal Intrauterine Undernutrition: A Comparison with Sibships," *American Journal of Epidemiology* 146, no. 10 (November 1997): 810–19.

283 **the "14-day rule":** Guidelines released by the International Society for Stem Cell Research (ISSCR) in 2021 suggest a path whereby researchers could seek regulatory approval through an independent review process to culture embryos beyond this point (*Guidelines for Stem Cell Research and Clinical Translation*, https://www.isscr.org/guidelines).

283 **cells autonomously self-assemble:** Leqian Yu, Yulei Wei, Jialei Duan, et al., "Blastocyst-Like Structures Generated from Human Pluripotent Stem Cells," *Nature* 591 (March 2021): 620–26; Xiaodong Liu, Jia Ping Tan, Jan Schroder, et al., "Modelling Human Blastocysts by Reprogramming Fibroblasts into iBlastoids," *Nature* 591 (March 2021): 627–32; Harunobu Kagawa, Alok Javali, Heidar Heidari Khoei, et al., "Human Blastoids Model Blastocyst Development and Implantation," *Nature* 601 (January 2022): 600–605.

284 **It is powerful, it is fast, and it is easy:** Doudna and Charpentier received the Nobel Prize in Chemistry in 2020 for their work on CRISPR. In addition to its applications in biology and medicine described in the text, CRISPR has also had a major impact on the agricultural industry, where it is being used to develop crops and livestock that resist disease and/or produce greater yields—see Haocheng Zhu, Chao Li, and Caixia Gao, "Applications of CRISPR-Cas in Agriculture and Plant Biology," *Nature Reviews Molecular Cell Biology* 21 (September 2020): 661–77. While the technical details of CRISPR are beyond the scope of this book, there are many excellent sources that describe the technique's discovery and its potential uses (and misuses), including an account by Dr. Doudna herself—see Jennifer Doudna and Samuel Sternberg, *A Crack in Creation: Gene Editing and the Unthinkable Power to Control Evolution* (New York: Mariner Books, 2017).

284 **or eliminate it altogether:** Researchers are taking at least two approaches to correct the disorder. The first uses CRISPR to correct the genetic defect itself—a gene-editing approach that converts the mutant version of the beta-globin gene to its wild-type version. The second involves deleting a gene whose product inhibits the production of an alternative form of hemoglobin—"fetal hemoglobin"—that could also ameliorate disease. To date, the latter approach has had the most success, as reflected in the reference below.

284 **freedom from the agony:** Haydar Frangoul, David Altshuler, Dominica Cappellini, et al., "CRISPR-Cas9 Gene Editing for Sickle Cell Disease and b-Thalassemia," *New England Journal of Medicine* 384 (January 2021): 252–60; Rob Stein, "First Sickle Cell Patient Treated with CRISPR Gene-Editing Still Thriving," NPR, December 31, 2021.

284 **other diseases are following close behind:** You Lu, Jianxin Xue, Tao Deng, et al., "Safety and Feasibility of CRISPR-Edited T Cells in Patients with Refractory Non-Small-Cell Lung Cancer," *Nature Medicine* 26, no. 5 (May 2020): 732–40; Morgan

Maeder, Michael Stefanidakis, Christopher Wilson, et al., "Development of a Gene-Editing Approach to Restore Vision Loss in Leber Congenital Amaurosis Type 10," *Nature Medicine* 25, no. 2 (February 2019): 229–33.

285 **several barriers have stood in the way:** Among the other concerns surrounding xenotransplantation has been the existence of porcine endogenous retroviruses (PERVs)—viral sequences embedded in the pig genome that pose a potential threat to the transplant recipient. Consequently, independent efforts are also underway using CRISPR to create genetically altered pig strains lacking all 62 known PERV sequences.

285 **traces of a pig virus:** Antonio Regalado, "The Gene-Edited Pig Heart Given to a Dying Patient Was Infected with a Pig Virus," *Technology Review*, May 4, 2022.

287 **the "geep":** Carole Fehilly, S. M. Willadsen, and Elizabeth Tucker, "Interspecific Chimeaerism between Sheep and Goat," *Nature* 307 (February 1984): 634–36.

287 **monkey-human chimeric embryos:** Tao Tan, Jun Wu, Chenyang Si, et al., "Chimeric Contribution of Human Extended Pluripotent Stem Cells to Monkey Embryos *ex Vivo*," *Cell* 184, no. 8 (April 2021): 2020–32.

287 **transgenic offspring:** Lei Shi, Xin Luo, Jin Jiang, et al., "Transgenic Rhesus Monkeys Carrying the Human *MCPH1* Gene Copies Show Human-Like Neoteny of Brain Development," *National Science Review* 6, no. 3 (May 2019): 480–93.

287 **Ethical questions have been raised:** Antonio Regalado, "Chinese Scientists Have Put Human Brain Genes in Monkeys—and Yes, They May Be Smarter," *Technology Review*, April 10, 2019.

287 **He defended himself:** Dennis Normile, "Researcher Who Created CRISPR Twins Defends His Work but Leaves Many Questions Unanswered," *Science*, November 28, 2018; Sharon Begley, "Amid Uproar, Chinese Scientist Defends Creating Gene-Edited Babies," *STAT*, November 28, 2018. He announced his results via YouTube on November 25, 2018, https://www.youtube.com/watch?v=th0vnOmFltc.

288 **immune to HIV infection:** A small percentage of the world's population naturally lacks CCR5, which has redundant functions in the immune system. Such individuals are naturally resistant to HIV infection.

289 **the physicist Robert Oppenheimer:** Oppenheimer's speech, "The Tree of Knowledge," was published in *Harper's* in October 1958 and reprinted with permission in *The Scientist vs. the Humanist*, edited by George Levine and Owen Thomas (Binghamton, NY: W. W. Norton, 1963).

290 **chance favors the prepared mind:** Louis Pasteur speech at the University of Lille, December 7, 1854.

291 **50 percent of the biological sciences workforce:** U.S. Bureau of Labor Statistics, "Employed Persons by Detailed Occupation, Sex, Race, and Hispanic or Latino Ethnicity," Labor Force Statistics from the Current Population Survey, Table 11, 2021, https://www.bls.gov/cps/cpsaat11.htm.

292 **can trace their origins:** The number of drugs directly attributable to pure basic science is hard to measure accurately. But several studies have examined the relationship between FDA-approved drugs and public sector funding (mostly NIH), confirming the strong links between basic research and new therapies. See, for example, Ekaterina Galkina Cleary, Jennifer Beierlein, Navleen Surjit Khanuja, Laura McNamee, and Fred Ledley, "Contribution of NIH Funding to New Drug Approvals 2010–2016," *Proceedings of the National Academy of Sciences USA* 115, no. 10 (March 2018): 2329–34; and Iain Cockburn and Rebecca Henderson, "Pub-

licly Funded Science and the Productivity of the Pharmaceutical Industry," in *Innovation Policy and the Economy* (Cambridge: MIT Press, 2001), 1–34.

EPILOGUE: PARTURITION

297 **the study of embryonic development:** Scott Gilbert, "Developmental Biology, the Stem Cell of Biological Disciplines," *PloS Biology* 15, no. 12 (December 2017): e2003691.

299 **"a profound source of spirituality":** Carl Sagan, *The Demon-Haunted World* (New York: Random House, 1995).

299 **"In a hundred billion galaxies":** Carl Sagan, *Cosmos* (New York: Random House, 1980).

INDEX

Page numbers in *italics* refer to illustrations.